Nature, Landscape, and Building for Sustainability

Harvard Design Magazine Readers
William S. Saunders, Editor

1. Commodification and Spectacle in Architecture
2. Sprawl and Suburbia
3. Urban Planning Today
4. Judging Architectural Value
5. The New Architectural Pragmatism
6. Nature, Landscape, and Building for Sustainability

Nature, Landscape, and Building for Sustainability

A Harvard Design Magazine Reader

Introduction by Robert L. Thayer Jr.

William S. Saunders, Editor

University of Minnesota Press | Minneapolis | London

The essays in this book previously appeared in *Harvard Design Magazine,* Harvard University Graduate School of Design; Peter G. Rowe, Dean, 1992–2004; Alan Altshuler, Dean, 2005–2007; Mohsen Mostafavi, Dean, 2008–.

Thanks to coordinator Meghan Ryan for her work on *Harvard Design Magazine.*

Excerpt from "Place," by Robert Pack, reprinted with permission of Robert Pack.

Every effort has been made to obtain permission to reproduce the illustrations in this book. If any acknowledgment has not been included, we encourage copyright holders to notify the publisher.

Published by the University of Minnesota Press
111 Third Avenue South, Suite 290
Minneapolis, MN 55401-2520
http://www.upress.umn.edu

Library of Congress Cataloging-in-Publication Data

Nature, landscape, and building for sustainability : a Harvard Design Magazine reader / William S. Saunders, editor ; introduction by Robert L. Thayer Jr.
    p.   cm. — (Harvard Design Magazine Readers ; 6)
    ISBN 978-0-8166-5358-4 (hc : alk. paper) —
    ISBN 978-0-8166-5359-1 (pb : alk. paper)
    1. Ecological landscape design.   I. Saunders, William S.
    SB472.45.N38 2008
    712—dc22
                                                            2007050743

Printed in the United States of America on acid-free paper

The University of Minnesota is an equal-opportunity educator and employer.

15 14 13 12 11 10 09 08   10 9 8 7 6 5 4 3 2 1

Contents

Introduction
Robert L. Thayer Jr.

Humans have torn themselves from the rest of nature, and sustainable design is the only way to repair the rift.

If it were only possible to summarize the thrust of this book in such a simple sentence, there would be no book, and I would not be writing its introduction. The interacting notions of nature, landscape, and sustaining design at times might seem simple, but they often slip sideways, like a blob of mercury, when pinned down. Historian Clarence Glacken took over seven hundred pages to trace the historic roots of nature and culture in Western thought in his classic text *Traces on the Rhodian Shore: Nature and Culture in Western Thought from Ancient Times to the End of the Eighteenth Century*.[1] Surely any possible resolution of humans' conflict with nature through an approach to life-sustaining design could not fully be illuminated in fifteen short essays, could it?

On the other hand, the path through complexity is sometimes elegantly simple. Buddhist masters speak of Four Noble Truths: the existence of suffering, the cause of suffering, the possibility of ending suffering, and a means to end it. Likewise, this book recognizes a human rift with nature, strives to understand its cause, contemplates a resolution, and offers meaningful steps toward reconciliation. Through this simple analogy we gain access to a book by some of the best thinkers on this postmodern predicament, a predicament rapidly rising to the surface of human problems. Where, and how, did we lose our way in nature? Are we to find a way "home" again? How will we do this? Where do we begin?

Readers of this book will begin with the words of philosopher Albert Borgmann, who, in an echo of Glacken's lengthy explorations on the Judeo-Christian theological construction of mystical nature, asks how we might return to that kind of premodern piety after modern technology has "left its imprint on everything" and become a kind of de facto operative religion with an undeveloped credo to accompany it. The result? In Borgmann's words, a postmodern divide that, to surmount, we must eventually stop seeking a state of grace in technology's "hyperreal elsewhere" and come to settle down in a traditional and familiar place.

But how can we do this, intones next author Bill McKibben, if humans have "supplanted God," and the resultant economy seems more real to us than the ecosphere in which, as economist Herman Daly reminds us, that economy is actually embedded?[2] Why do we not see the human economy as embedded in the ecosphere, but, instead, are overwhelmed by our most synthetic human constructs? "Artificial is human made, humans are natural, artificial is natural," reads the small anonymous quote in the $3 booklet I bought in a San Francisco bookstore during the height of the social upheaval of the 1960s. The economy and its driving technologies are indeed "artificial," but we humans are part of nature. What gives? Quite simply, *scale*.

The physical landscape's changes wrought when we humans began to use fossil fuels soon dwarfed those produced by human or animal power alone. The immense releases of energy from machines burning first coal, then oil and gas, then from those using electric power and nuclear fission simply overwhelmed the landscape and converted it to something entirely beyond its former scale. On the ground and in the mind, the scale of landscape exploded, and human impacts reached around the globe.

Human-felt alienation perhaps began with that scalar explosion. Landscape architect Sir Jeffrey Jellicoe once wrote, "Ugliness in works of man began when there became a conflict between what lay on the land's surface and what lay beneath." Can we grow or harvest it? Or must it be mined? These are completely different concepts, universes apart in terms of space, time, land, and scale. Throw in the relative invisibility and imperceivability of genetic manipulation, nanotechnology, and nuclear fission, and it is no wonder that McKibben laments the "magnitude of the changes now under way," noting that "the irony, perhaps, is that our domination may carry the seeds of our own diminution."

In the next chapter, however, Lucy Lippard reminds us, in "Too Much: The Grand Canyon(s)," that it is not always human physical construction or elaborate technology that seems to explode perceptual scale and overwhelms our sturdy relation to nature, but the ways in which our postmodern culture assigns meaning to things, concepts, and places. Speaking succinctly about our reactions to the country's preeminent natural landscape, she speaks of the canyon in tandem terms: as both an accumulation of differing human sentiment and meaning, and a presence beyond human comprehension. The place is "off the charts," she says, reminding us that layer upon layer of postmodern meaning accrues to the Grand Canyon as if deposited like the sedimentary rock incised by the canyon itself. Yet it is off the charts not just because the human culture that deposits, interprets, and disseminates its various meanings, like neighboring Las Vegas, is itself beyond simple comprehension. "Is Thoreau's 'vital and intense sense of connection to nature' possible in a skeptical postmodern era?" she asks. Speaking of the grandeur of nature, she writes, "anyone who can avoid a certain romanticism in the Grand Canyon is probably hopeless." In spite of the abundance of assigned meanings, the overwhelming power of the canyon is still a force for personal transcendence.

Writing, like the previous authors, in a pre-9/11 context, professor emerita Catherine Howett ponders the "sense of helplessness about our capacity to protect ourselves or the people and places we care about, much less the human race and the planet." As if the anomie of the postmodern world of 2000 were not enough, events since 2001 have only heightened that sense of helplessness and added additional dimensions, as the American twin towers of shared democracy and world respect have tumbled. Add the hyperbolic revelations of global climate change, and it is no wonder that contemporary society feels helpless and in need of a new kind of protection for that which we love. But as if to anticipate this, Howett tells us that to resolve this situation a "radical transformation of contemporary culture" must occur, echoing Robert Bellah's landmark 1985 work on the essentials of American society, *Habits of the Heart*.[3] But Howett reminds us also that to accomplish such a transformation, the design arts must reach "beyond the educated elites to engage ordinary men and women." Lucy Lippard would wholeheartedly agree.

Such a transformation to a more sustaining way of living, designing, and building must first surmount a critical hurdle: the commodification of nature. Critic John Beardsley writes in his essay "Kiss Nature Goodbye: Marketing the Great Outdoors" that the nature presented in chain stores like the Nature Company and REI is "not localized, dynamic, and differentiated into various ecosystems but homogenized, static, portable, and consumable." The great "middle landscape" of Carl Sauer and Leo Marx is where animals and other objects of nature in the zone between wilderness and city might even be beneficial to the preservation of native habitats in light of the placelessness generated by commodified nature in hyperreal, themed environments.[4] If we can build convincing fakes, will we continue to value the original? The deliberately themed construction of "natural" environments lulls us into the illusion of being able to consume while we "conserve." Beardsley answers a critical rhetorical question of his own making and resolves the dilemma described above: "Can we imagine a theme park that is genuinely fun and truly educational and environmentally responsible all at once? I don't see why not." Indeed, it would seem to be an imperative step.

Michael Pollan contrasts the two great polarities in the American landscape tradition: lawn and wilderness. He is quite correct to assume that the garden is a microcosm of much larger ideas about nature. The front lawn, that uniquely American manifestation of the English manor estate, opposes in spirit, if not in fact, the transcendentalist (and of late, arguable) idea of a wild, raw, unpeopled state of nature. The latter idea is perhaps the singular identifier most capable of separating American from continental European culture, yet it is the former, lawn, which most wistfully reminds us of our English landscape heritage. These traditions, while apparently polar opposites, actually begin to meld, as researchers like Kat Anderson remind us that what we called wilderness is not unpeopled at all, having been modified deliberately by Native Americans for centuries.[5] Likewise, current trends in the front gardens of American suburbia frequently bend toward native plants and ecosystems from local bioregions. So, as Pollan concludes, the continuing evolution of American garden tradition offers a "third way" into the landscape. Manifestations of stewardship implicit in the concern for a sustainable future will certainly bear him out.

At the book's fulcrum, we are reminded by Rossana Vaccarino that although Gifford Pinchot's original conservation and "wise use" ethic was later co-opted by the very industries he wished to keep at bay, the genesis of "sustainability" arose from the nascent notion of sustained yield in early scientific American forestry, something that few students of "sustainability" today recognize. History, in many ways, is never static, always revisionist. Among the historical constructs now in flux affecting the topic of this book are that nature had to be rhetorically "emptied" of more than ten thousand years of indigenous human occupation in order to be "occupied" (by white men) for the preservation of "nature"—a brilliant interpretation supported by the work of ethnobotanist M. Kat Anderson in her book *Tending the Wild: Native American Knowledge and the Management of California's Natural Resources.*[6] Such interpretation by Vaccarino, Anderson, and others throws a curve ball at our typical interpretation that vast empty wilderness (and, hence, the wilderness-derived American myth of rugged individualism set against a tabula rasa of wild nature) is to continue to be the foundation of American self-consciousness. To continue such illusions unchallenged would leave little room for the explicit reintegration of nature and culture through the sustaining design championed by this book.

Susannah Hagan, in "Five Reasons to Adopt Environmental Design," reminds us that reappraisal of material culture is late to arrive at architecture, and we should not confuse rhetoric with substance ("'It is not language that has a hole in its ozone layer'"). Reading Hagan, we are taught that nature, rather than being a set of fixed objects and impressions, is dynamic, complex, both very large and very small, and that architects (and landscape architects, too) have found it difficult to capture its dynamism and to endure many of the same constraints of building within the Newtonian subset of an increasingly mysterious, post-Einsteinian cosmology. Both professions are understandably impatient for the means of incorporating the "new" nature of nature into the "old" methods of design. Is there a hint here of what we will read about in the next several chapters?

Architect Peter Buchanan, author of "Invitation to the Dance," has worked all over the world and is well qualified to deliver what amounts to a basic primer on green architectural design. Europe has led the world in operationalizing much of what America has only been talking and writing about, as evidenced by the Beddington Zero Emission Development in the United Kingdom. The project reduces the ecological footprint per resident (another practical heuristic device introduced to readers by Buchanan) to a meager 2.2 hectares per person, instead of the UK average of 17.3. His essay is now taking us from the theoretical to the very practical, yet back again: "This is the big choice we face: to move from the ego to the eco; from acting on the world to acting *with* it; from standing alone like Howard Roark to joining the dance." Clearly, there is a pathway revealing itself here, with the first stepping-stones crisply in focus.

But our destination is not free of hurdles en route. Landscape architect Robert France asks the rhetorical question of whether the gap between promise and the reality of landscape architecture can be spanned? The answer is a qualified "yes," illustrated by examples of wetland parks and gardens that convey both the feelings and the function of intact natural systems. As perhaps the most highly visible, expressive, and atavistically valued component in nature, it makes sense that *water* illustrates the very pathway design must take in sustaining life and the human spirit. But we must also consider infrastructure, energy, and light: Kristina Hill, in "Green Good, Better, and Best," addresses what many (including the writer of this introduction) consider to be the crux of sustaining design: civic infrastructure. Comparing situations in Berlin and Seattle, Hill takes France's theme even further, concluding that a collection of mere "demonstration projects" does not a sustainable world make; a much deeper and more systemic, inclusive approach is called for, one that will succeed only when a new, sustaining urban ecosystem is born. The late John Tillman Lyle, author of *Regenerative Design for Sustainable Development,* would approve.[7]

Which brings us to "Energy, Body, Building." This essay arrives logically at the place where all human "environments" begin: the skin, the eyes, and other sensory organs. Are architects ignoring the physiology of the human body to fulfill some outdated notions of practical building? Should top-down, politically interpreted energy conservation norms be allowed to ignore the fact that the eyes see light and the body feels heat or cold *relatively*? Author, engineer, and architect Michelle Addington compels us to base design practice on the human as a living organism, not as a bureaucratic automaton. If we are able to do so, we discover the true efficiencies embedded in the nature of humans, efficiencies that might translate directly into real reductions in per capita energy for lighting, heating, and cooling.

Niall Kirkwood offers a fascinating account of phytoremediation, that mouthful of a word that refers to vegetation systemically absorbing toxins that have entered the landscape as by-products of human industrial processes. By describing the biological systems that may control the excesses of industrial infrastructure, Kirkwood unwraps a corner of a much larger frame. It is not a long stretch of imagination from his chapter to a future in which ecological principles have thoroughly transformed much of our former industrial systems and, as Lyle predicted, human ecosystems will manifest the same multifunctional regenerative capacities as natural ecosystems.

But wait a minute. Before we imagine too fancy an ecological future, we are admonished by Peter Del Tredici, in "Neocreationism and the Illusion of Ecological Restoration," to take some simple advice: "Don't limit your planting designs to a palette of native species that might once have grown on the site. Imposing such a limitation on diversity not only reduces the aesthetic possibilities for the landscape, but also its overall adaptability. As a graceful way out of the native versus exotic debate, I recommend using sustainability as the standard for deciding what to plant." No sustainable future can emerge by returning to the past; evolution never gets us back to the same place twice. The mere invasiveness of plants (like their migratory human counterparts) must ultimately be judged on their effects on the whole landscape. And herein lies the book's most important lesson: the obvious constraints and mistakes of objectifying or commodifying nature yield to a dynamic, systemic model of the future, where the intent is to achieve in the human realm the complex efficiencies, yet simple elegance recognizable in design of natural systems.

In "A Word for Landscape Architecture," art historian and theoretician John Beardsley summarizes the acceleration of theoretical exploration and the rapidly increasing complexity of landscape architectural works. It is a fitting finale to the book. Beardsley's bold assertion of a positive future for both the artistry and environmental relevance of landscape architecture echoes the themes presented in previous chapters, and yet still moves beyond them. Although *entropy* is perhaps the word toward which Beardsley points this final chapter, he is perhaps more accurately aiming at *complexity* or the term *negentropy* of physicists like Erwin Schrödinger and information scientists like Norbert Wiener. It is actually the retardation or slowing down of the entropic process, by which the landscapes

Beardsley describes build higher and higher levels of order, complexity, sophistication, and, therefore, sustainability. I must say that I, too, am convinced of the continuous sunrise, rather than purported sunset, of landscape architecture as it further engages and reveals the dynamic unfolding of the world's living systems. No other artistic medium is so dynamic, yet so essential, to life on earth.

So, what might readers learn from this book? That the divisions between nature and culture are complex, dynamic, and socially constructed; that godlike properties interconnect them; that the *scale* of human intervention has moved beyond our perceived ability to comprehend and adapt; that there are both conflicts and congruencies between how we *feel* about nature and what we *do* about it; that nature has been commodified and sold back to us for a profit. But reading this book might also reveal that the term *sustainability*, in spite of its slippery ubiquity, is grounded in the best of American conservation tradition; that it makes ultimate good sense as an armature for future design practice; that the effectiveness of sustainable design depends on moving away from endless "demonstration" projects into the viable mainstream; that *human* nature requires designing sustaining environments from the body out as well as the universe in; that the gap between the promise and the reality of sustainable design, although significant, can be breached by diligence and, most of all, by *practice*.

Michelle Addington perhaps summarizes best the reasons designers need this book when she writes, "Let us, however, think in territories larger and broader than the things we make."

Notes

1. Clarence Glacken, *Traces on the Rhodian Shore: Nature and Culture in Western Thought from Ancient Times to the End of the Eighteenth Century,* reprint ed. (Berkeley: University of California Press, 1976).

2. See Herman Daly, *Beyond Growth: The Economics of Sustainable Development* (Boston: Beacon Press, 1996).

3. Robert N. Bellah, Richard Madsen, William M. Sullivan, and Ann Swidler, *Habits of the Heart: Individualism and Commitment in American Life* (Berkeley: University of California Press, 1996).

4. See Carl Sauer, *Land and Life : A Selection from the Writings of Carl Ortwin Sauer* (Berkeley: University of California Press, 1974); and Leo

Marx, *The Machine in the Garden: Technology and the Pastoral Ideal in America* (New York: Oxford University Press, 2000).

5. M. Kat Anderson, *Tending the Wild: Native American Knowledge and the Management of California's Natural Resources* (Berkeley: University of California Press, 2005).

6. Ibid.

7. John Tillman Lyle, *Regenerative Design for Sustainable Development* (New York: Wiley, 1994).

I | Imagining Nature

1

The Destitution of Space: From Cosmic Order to Cyber Disorientation

Albert Borgmann

I'd laugh out loud because
just being in no special place,
yet being there alive with time to pause

and words upon my tongue
enabling me to measure out that pause as such
and thus possess my pausing self
(as if my mind could touch

its wild reflection in the splattered shade)
seemed all the cause laughter required.
—Robert Pack, from "Place"

The great Gothic tower of Freiburg Minster is widely visible in this city in the Upper Rhine valley. At 372 feet it is nearly three times the height of surrounding buildings. But distant acquaintance with the cathedral leaves you unprepared for the grandeur encountered up close. As you approach the tower from the west and at a slight angle to its facade, an enormous portal gradually appears. The portal is set within the massive base of the tower, whose shaft becomes a strongly articulated octagon that ends with the filigree of the steeple. The force lines of the mountains to the east, of the buildings around the square, and of the very cobblestones beneath your feet all seem to be gathered and swept up by the tower to culminate in the cruciform stone flower and the star and moon that crown the spire. You sense that for those who created this work of architecture, the sky above was heaven.

As you pass through the portal into the vestibule, the great figures of the history of salvation—the patriarchs, prophets, and saints—accompany you to the tympanum of the doors that lead to the interior of the church. There you find the movement toward heaven represented in a relief depicting the Day of Judgment. The departed rise from their graves. The blessed ascend to the glory of heaven, the damned are dragged down into the jaws of hell.

The tower of the church at Freiburg, set between heaven and hell and imbued with the history of salvation, is the focal point of a great space, much as the Acropolis, hilltop realm of the gods, centered Athens within a cosmology that encompassed the Greek city-states and Mount Olympus. These, of course, are spaces of the past. The fourteenth-century cathedral and the ancient Greek temples no longer command our devotion. More to the point, they have had no successors in our time. Contemporary space is, I want to argue, a space of destitution. It has lost its cosmic scope and its terrestrial orientation; most radically of all, it has ceased to be the depth-conferring space of memory and imagination.

The history of cosmology reveals an unmistakable pattern. The earliest, mythic accounts of the universe depict a space inseparably spiritual and material; the picture joins the world's timbers and a divine order that assigns humans their place and obligations. The Hebrew Book of Genesis tells us that the world is God's handiwork—well ordered and good. The cosmos conceptualized by Aristotle has no Creator, but it too is well ordered, with humans ranking above minerals, plants, and animals but below the divine intelligences (the planets and stars) and the entity Aristotle named the "prime mover."

These worlds were understood to be inhabitable. They had the earth for a fundament and the sky for a shelter; they were thus attuned to the vertical axis of human bodies. The first crack in this conception of a fully habitable world appeared in Western culture when, in ancient Greece, the astronomer and mathematician Eudoxus hypothesized that the earth was a sphere. How could people live on the other side of a circular earth and not fall into the cosmic abyss?

Even as early as Ptolemy, in the second century AD, the relationship of cosmic structure to the divine order was seen as an explicit problem. Ptolemy carefully, not to say precariously, placed astronomy between the tangible, changeable realm of physics and the sublime, lofty realm of theology. Thirteen centuries later, Copernicus, when he proposed the heliocentric revolution, still felt he had to justify the central position of the sun theologically. "In the middle of all this sits Sun enthroned. In this most beautiful temple could we place this luminary in any better position from which he can illuminate the whole at once? He is rightly called the Lamp, the Mind, the Ruler of the Universe; Hermes Trismegistus names him the Visible God, Sophocles' Electra calls him the All-seeing."[1]

In Isaac Newton's universe, divinity was no longer present; it had left only the trace required to explain those features that did not follow from the natural laws Newton propounded; for example, the motions of the planets in the same plane and direction "could not spring from any natural cause alone, but were impressed by an intelligent Agent."[2] As modern astronomy and cosmology extended their explanatory scope over the next three centuries, the need to invoke a higher being or divine intelligence evaporated. Albert Einstein's references to God as "the Old One" are for the most part endearingly anachronistic metaphors.

Today the universe is uninhabitable for human imagination. The first thing you teach in contemporary cosmology is that space is isotropic: it has no privileged center or direction. Students will accept this obediently and think of our galaxy as traversing a space that is similar in all directions. But this understanding is severely tested when they are asked to contemplate the beginning of the universe. As the physicist Steven Weinberg has written, "In the beginning was an explosion. Not an explosion like those familiar on earth, starting from a center and spreading out to engulf more and more of the circumambient air, but an explosion which occurred simultaneously everywhere, filling all space from the beginning, with every particle of matter rushing apart from every other particle."[3] There goes our picture of the cosmic pea that expands into a colossal universe, and there too goes whatever imaginative grip we had on the Big Bang.

This is not to say that contemporary cosmology is unintelligible. One can learn to distinguish true from false cosmological propositions, and computation is a significant support in the learning process. What we cannot do is imagine ourselves witnessing the Big Bang, nor can we think of ourselves as inhabiting the resulting geometry at the particular point called earth. The cosmos has become scientifically comprehensible, yet imaginatively inconceivable and hence uninhabitable.

This spatial disorientation, however, is merely one aspect of what appears to be a deeper and in fact total lack of significance, be it theological, ethical, or aesthetic. Weinberg also wrote, "The more the universe seems comprehensible, the more it also seems pointless."[4] One might shrug off this uncomfortable conclusion. I am writing this essay in the United States, and I suspect that most readers of this essay live too in the developed world, amid comfort and abundance. We do not reside in cosmic space. But to be content to dwell only in our immediate surroundings reveals, perhaps, a confinement and impoverishment in our turn-of-the-millennium culture, which otherwise prides itself on its openness and richness. Premodern cultures were keenly conscious of space; we might call them spacious, in the sense of being "space-full." They could and did inhabit, at least conceptually, the whole universe. In contrast, the semantic space most people in our rich Western democracies inhabit is just the surface of the earth. The light-polluted night sky obscures the stars, planets, and galaxies beyond the earth; our insolent indifference banishes from awareness the cosmic structure of which our planet is a part. The earth is part of a solar system, and the sun, far from being the center Copernicus took it to be, is an average-sized star in a galaxy that in turn belongs to a cluster and supercluster of galaxies, and it is at the level of superclusters that the universe turns out to be pretty much featureless. Thus, we seem destined to live either in comprehensible pointlessness or with impertinent orientation.

Surprisingly perhaps, the pointlessness we try to banish to outer space has invaded and disoriented the proximate world as well. But what was the pristine spatial condition that has been so transformed? Navigation proceeds in two fundamentally distinguishable ways, by landmarks or by instruments. In western seafaring, going by landmarks meant hugging the shore. For the Blackfeet in the American West, mountain peaks and rivers were large-scale landmarks; buttes, trees, and cairns were more particular points of orientation.

The word *orientation,* of course, derives from the rising sun *(sol oriens)* and reminds us of the role heavenly bodies once played in pointing the way. We need to understand, however, that sacred sites, at least the eminent ones, were not just precursors of navigational technology, of instruments. Rather the great monuments served to establish something like an ordered space. For the Blackfeet, the Continental Divide in what is now Glacier National Park was the backbone of the world, the ridge that gave reality its shape and stability. Chief Mountain was the lofty point where Blackfeet boys on their vision quest learned to comprehend the world and their place in it.[5]

The eagerness of cartographers to depict the earth accurately and the need of sailors to make their way on trackless waters led to the displacement of the oriented space of landmarks with the technological space of grids. At the beginning of the modern era, once compasses and reliable maps were available, dead reckoning could determine a ship's location. That method used an implicit system of polar coordinates, reckoning a location according to the direction and distance of a ship from its point of origin. But holding the direction and determining the speed of a ship proved so elusive that the inevitable errors led often to hunger, disease, and death. The explicit coordinate system of latitudes and longitudes was until the late eighteenth century only half useful. Latitude could be determined by the angle of the sun at noon, but longitude remained impossible or difficult to calculate until reliable clocks indicated the precise difference in time at noon and so the distance in space from the principal meridian in Greenwich, England.[6]

Space appropriated through grids and instruments may be called *technological*. Technological space has little need for the eloquence of nature. Hence it replaces what was once understood to be the natural coherence of space with an abstract matrix or scaffolding in which natural places are arbitrarily, though controllably, located. The ideal mode of travel within technological space consists of a pod containing a chair, a table, and a computer. You go in, close the door, sit down, type in your destination, press enter, get up, open the door, and you are at your destination. Plane travel is an imperfect approximation of this ideal. You need to type in your destination days before you travel. The cabins are uncomfortable and inconvenient, and you have to share them with dozens or hundreds of people. Most of the time the execution of the enter command takes so long that you get hungry or bored and need to be fed or shown a movie. And what is more, for the air traveler in Missoula, Montana, the difference between Minneapolis and Salt Lake City is not that Missoula and Minneapolis are separated by the Continental Divide, the Rocky Mountains, the northern plains, and finally the verdant fields of the Midwest, whereas Missoula and Salt Lake City are separated by the Clark Fork valley, the Pioneer Mountains, and the Wasatch Range. The difference is simply that it is two hours to Minneapolis but only one to Salt Lake City.

Ironically, the most radical displacement of natural by technological space bedevils the very enterprise designed to give us the most comprehensive and penetrating view of the earth: geographical information systems, or GIS. Typically, GIS data are emitted by reflections of solar or satellite radiation; the information is sensed by satellites, and, once processed, displayed on a computer screen.[7] Considered scientifically and technologically, GIS data are astounding and admirable in their amount and variety. One irony, however, is that GIS removes biologists, geologists, geographers, and ecologists from actual contact with rocks, plants, and animals—from the endless and specific intricacy of nature—and shackles them to a chair in front of a monitor. The scientist's time is too valuable to be frittered away on the ground. Another irony is that the connections between information and reality are left to be ascertained by "field grunts," whose work is considered both tedious and expensive and is often, in fact, so thin that "ground truthing," as it is wistfully called, can become dangerously tenuous, while incredible amounts of ambiguous data pour into and out of computers.

Coordinate systems in the wider sense of sets of spatially ordered signs have invaded and transformed not only the grand sweep of our worldview but also the mundane details of life. Documents are stored in filing cabinets ordered top to bottom and front to back. When you visit a new acquaintance in a metropolitan area, you find your way by the number and direction of an expressway, the name or number of an exit, the number of blocks first in this direction and then that, and finally by a house number. When you go to see your lawyer in her downtown office, you are given the location of a high-rise, the floor, and the number of the office.

Technological space comes in different degrees of purity. Determining your location by map and compass requires a fair intimacy with the lay of the land. Determining it through a global positioning system (GPS) requires no such intimacy; indeed, it places you in the world with the brute disconnectedness of arrival by plane. Conversely, you move from a more purely technological space to one more traditionally oriented when you move to a new job and city. At first, the values of some coordinate system are the devices that help you find home and office. But gradually a corridor of familiarity and orientation emerges and makes careful watching and counting unnecessary. In time the corridor widens into some intimacy with the city, its landmarks and environs. Still, for trips to unfamiliar neighborhoods or distant places, the abstract account of a map and some technological device of location and transportation delimit your spatial competence.

A space whose rigidity determines locations so that the distances between them can be measured is said to have a metric. A space wherein relations and continuities alone matter is, informally speaking, topological. Physical space is unyieldingly metric, and at one time humans paid tribute to the rigor of its distances by enduring the two weeks or two months of an Atlantic crossing or by traversing the Bozeman trail. These kinds of distances have now disappeared into mere degrees of inconvenience and the difference of a few hours of air travel.

The perfection of technological space is reached in the measureless ubiquity of cyberspace. Here the last traces of metric space have disappeared.[8] A precursor of distanceless space is the telephone system. By the mid-twentieth century it had leveled the distances within the area of local calls. Distances to other cities and continents were still marked by the interventions of the long-distance operator and by significant expense. Moreover, the entry points to the system were few. Many a customer, excitedly summoned to a long-distance call, arrived breathlessly at one of the widely scattered telephones. Today we can call and be called from anywhere, at any time.

The telephone system exemplifies the norm of easy and immeasurable distancelessness that characterizes cyberspace. But it is a thin slice of reality that is thus accessible. Making most of reality—in all its color, sound, and motion—available with the ease and ubiquity of telephoning is the great goal of information technology.

There is a conceit in contemporary culture that the perfection of cyberspace will in some part supersede and obviate air travel. A cubicle with a computer connected to cyberspace, so the argument goes, will transport us instantly wherever we want to be or, what comes to the same thing in distanceless space, will in seconds deliver to us, in some form, whatever piece of reality we desire. This shift from the orientation and metric of natural space to the isotropic and topological space of technology and information would seem to be disorienting and unsettling. But people new to this technology grasp the essence of technological travel and the pleasures of television and the Web within hours, if not minutes. Why this easy familiarity? Humans have always been intimate with ametric space—with the space of memory and imagination. I can recall distant places as quickly and easily as the space outside my office where I am now writing. My memory of Oahu calls up that house in Kaimuki as readily as my nostalgia for Freiburg conjures up that house in Herdern.

Dangerously, however, this kinship of mental and memorial space not only gives you ready access to cyberspace, it also allows the seductive explicitness and glamorous richness of cyberspace to invade and subdue imagination and memory. Internet industry and commerce are employing the most powerful talents to displace the life of your imagination with entertainment and advertising. It is the professed goal of America Online "to lure users to stay on line as long as three hours a day—the current average is 55 minutes. . . ."[9] As for memory, its exercise is derided by proponents of electronic education whose discourse invariably yokes "memorizing" together with "rote" or "mindless." You need to know or remember nothing, they imply, as long as you can access information on the Internet.[10]

Just as technological space has not entirely dissolved life in oriented space, however, so cyberspace will never fully colonize memory and imagination. We still see our adult children's endeavors as the mature exercise of their youthful talents and infant traits. We recognize how much more deeply Grand Central Station reaches into the past than does the Mall of America. But our world has little of the depth and eloquence that Athenians experienced with the help of their writers and orators. The firm and spacious structure we associate with a metric space (along with memory and imagination) has been lost in cyberspace and has given way to the pointless availability of anything and everything.

Premodern space was constituted according to the location of sacred monuments through which we oriented ourselves. It was understood to be enveloped by a numinous cosmic and memorial space. Of course, it was not always and entirely so; superstition, ignorance, and prejudice all too often clouded this space. We like to feel superior to such failings and yet also to be entirely unencumbered by norms of spacious plenitude. The destitution of contemporary space has left us with a world whose cosmic extension is thought to be irrelevant, whose global technological structure is disorienting, and whose imaginative and memorial depth has collapsed into indifference. An uninhabitable world, indeed.

At the same time our contemporary developed world is uniquely and enormously affluent. In fact, the character of contemporary affluence is of a piece with the destitution of space, for affluence today consists in the enjoyment of ubiquitous and refined consumer goods.[11] For a life of affluent consumption, contemporary cosmology is a useless headache, monuments of orientation are anachronistic oppressions, and the exercise of imagination and memory is a forbiddingly austere enterprise. We are largely unaware of the destitution of space, because we think that the threat to space is material poverty rather than experiential destitution. We recognize and reject, as well we might, a poor space, one that is confined, confusing, or terrifying. Having overcome the threat of poverty, we mistakenly think that all is well with the space we know.

Sometimes the rejection of premodern space goes beyond the unreflective enjoyment of affluent consumption and takes the form of triumphant postmodern theory. The leading idea is that to ignore or reject cosmology is to overthrow the master narrative and claim one's autonomy at last and fully. In this view, to be everywhere and nowhere is to be cosmopolitan. To be rid of the burdens of measure and memory and to assume polymorphous roles is to gain the promiscuous pleasures of diversity. Deconstructive postmodernism depends on what it opposes and slowly devours. But this much must be granted: the reconstitution of premodern space is impossible. If there is a constructive postmodern response to the destitution of space, it has to be a kind of construction. But with due respect to the various constructivists, we need to realize that construction is not a creation ex nihilo, but an enterprise that needs to discover designs, materials, and a ground to build on.

A return to premodern piety is impossible because modern technology has left its imprint on everything. If the holy is the wholly other, nothing holy is left to worship. There is a heated debate among environmental philosophers as to whether ecological restoration violates the autonomy of nature.[12] But this dispute overlooks the fact that untouched nature no longer exists. Through climate change, weed management, endangered species legislation, fire regimes, and much more, we are modifying nature inevitably everywhere and all the time.[13] The only question that remains is *how* we do this. Insofar as this is conceded, however, nothing yet follows for or against the destitution of space. Technology does not have to be self-destructive or unjust, and hence the environmental or liberal arguments we like to use in airing our discontents do not avail. We must face up to the aesthetic, moral, and metaphysical questions that would remain even if our ecological and social complaints were fully answered.

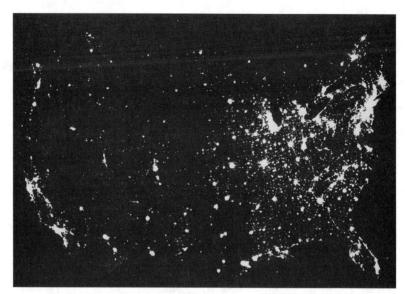

Light pollution over the United States as seen from space. Courtesy of the International Dark-Sky Association and Woody Sullivan.

To begin with aesthetics—works of art require a measured space. Size, symmetry, proportion, and rhythm dissolve in a purely topological space. We can regain a metric space through deliberately having *a*technological experiences—getting off our snowmobiles, jet skis, all-terrain vehicles, out of our powerboats and automobiles. Of course, we cannot totally "demobilize." Rather, demobilization should restore the nearness and openness of the focal areas of life, the treasured urban and natural spaces. Environmentalists who oppose mechanized travel in natural areas and New Urbanists who build walkable neighborhoods understand this well. Not surprisingly, these champions of measured space show an affinity as well to the metric of time that is marked by historical forms and processes. But important as these endeavors are, they do not by themselves yield grand spaces.

Premodern orientation through worship at the altars of sacred monuments may be a species, but it cannot be the model, of postmodern orientation. As the end of nature has given us a new kind of responsibility and enforced a new kind of maturity, so in the case of orientation something more productive and responsible than a (once appropriate) receptive and childlike piety is required. Focal points of

orientation will emerge through the correspondence of human cele-
brations and real landmarks. Some recently constructed baseball parks
come to mind. In contrast, the landmarks of nature require devotion
(since, for the most part, they cannot or should not be constructed)
to provide orientation. Cross-country skiers, canoeists, and hikers
meet snowmobiles, jet boats, and off-road vehicles with a sense of be-
trayal that leaves the users of these motorized conveyances puzzled.
Riders on dirt bikes are inclined to extend a friendly hello to the hikers they meet. But the hikers tend to remain reserved if not grim.[14]
Why these different responses? What for the hikers is the glorious far
point of a strenuous day is for the riders merely the halfway mark of
an afternoon outing. The old bikes used to take five hours to make
the loop; the new models need four hours; no doubt the next generation soon will be back home within three and a half hours. What
the hikers sense is the breach of ancient metric space that technology
inflicts on nature. The firmness of space along with its burdens and
blessings are no more, unless a majority of us agree to restore and
protect them.

Once measure and orientation have been restored to space through
demobilization and celebration, is there any hope of joining the perspectives of contemporary cosmology to the good of landmarks and
the beauty of measured space? Our society is not without comprehensive and much discussed assessments of its state of affairs. Books
such as *The Work of Nations, Earth in the Balance,* and *The Road
Ahead* propose to take the economic, environmental, and technological measure of our time, and they receive applause, criticism, and
elaboration.[15] But none of these proposals and discussions so much
as hints at a context spacious enough to pertain to the beginning
and end of all things. Is this a smug sort of ignorance? Or rather the
reasonable response to contemporary cosmology, which is, as I have
averred, an incomplete, hard to grasp, and spiritually silent theory?
However these questions will be answered, enough is now known to
upset profoundly our everyday notions of space, time, matter, and
energy. Should we conclude, as Steven Weinberg does, that we live in
"a tiny part of an overwhelmingly hostile universe"?[16] At the least,
this cosmology will rouse us from the slumber of our anthropocen-
trism and give us a more spacious awareness and humility. Today
familiarity with the universe must take the form of comprehension
rather than unmediated inhabitation.

We can glean something of this comprehension from the poems of Robert Pack, which ponder the wonders of the cosmos and engage the reader in its measures and rhythms.[17] In the lines that serve as the epigraph for this essay, Pack acknowledges the placelessness of the cosmos and is yet moved to laughter. What are we to make of his reaction? Weinberg, when faced with the incongruity of the human search for meaning and the pointlessness of the universe, found "some of the grace of tragedy" in the resolve, if not in the results, of research.[18] Another reaction to incongruity is possible, however, as theorists of laughter tell us. "For in disclosing dislocations everywhere, eventually in the cosmos," philosophers Gordon Brittan and David Healow have said, "humorous incongruities upset our worldviews, undermine our arrogance, and limit our responsibility."[19] But why is the cosmic incongruity doleful for Weinberg and humorous for Pack? The physicist's incongruity is lopsided, listing heavily toward the scientific structure of the universe and leaving us with a greatly attenuated human standpoint. The poet embodies the eloquence and exuberance of his location, and thus the arc of incongruity can grow to the point where laughter resolves it in a uniquely human way. Although the structure of the cosmos remains enigmatic, the pleasure and detachment of humor can engender a large-minded view of our place in the universe and hence, perhaps, a measure of magnanimity in our designs.

We live in ambiguous space. There are still the valleys and mountains of the wilderness, traces of adventure in air travel, vast expanses on the interstates, and cherished vistas and corners in our cities. For the most part, though, these are remnants. Many of the technological and economic advances often considered as evidence of our cultural vitality are smoothing and accelerating transportation, information, and consumption and are thus fomenting the feeling of restlessness and unreality that is the curse of destitute space. Complicity with technological space reduces design to brutality or insouciance—gigantic structures or arbitrary stylizing. Design is the discipline entrusted with the construction of space. It is the art that artisans and artists have in common—the marriage of craft and creativity. We know that creativity will not prosper without the teaching and exercise of craft. Part of the craft is to learn that a prosperous space needs to be measured, oriented, and magnanimous.

2000

Notes

1. Nicolaus Copernicus, "On the Revolutions of the Heavenly Spheres," in *Theories of the Universe,* ed. Milton K. Munitz (New York: Free Press, 1965), 169.

2. Isaac Newton, "Letter to Richard Bentley," in *Theories of the Universe,* ed. Munitz, 212. For a summary of how today we can explain what was inexplicable to Newton, see Philip Morrison, "The Hidden Cosmic Ruckus," *Scientific American,* July 1999, 104 and 106.

3. Steven Weinberg, *The First Three Minutes,* 2nd ed. (New York: Basic Books, 1993), 5.

4. Ibid., 154.

5. James Welch, *Fools Crow* (New York: Penguin, 1987), 3.

6. Dava Sobel, *Longitude* (New York: Penguin, 1996).

7. David J. Maguire, Michael F. Goodchild, and David W. Rhind, eds., *Geographical Information Systems* (Harlow: Longman, 1991).

8. More on the structure of cyberspace can be found in my "Information, Nearness, and Farness," in *The Robot in the Garden: Telerobotics and Telepistemology in the Age of the Internet,* ed. Ken Goldberg (Cambridge, Mass.: MIT Press, 2000).

9. Saul Hansell, "Now, AOL Everywhere," *New York Times,* July 4, 1999, section 3, 1.

10. More on "distance learning" can be found in my *Holding On to Reality* (Chicago: University Chicago Press, 1999), 203–8.

11. More on availability and consumption can be found in my *Technology and the Character of Contemporary Life* (Chicago: University of Chicago Press, 1984).

12. For the state of the discussion and pragmatic solution, see Andrew Light, "Ecological Restoration and the Culture of Nature: A Pragmatic Perspective," in *Restoring Nature: Perspectives from the Social Sciences and Humanities,* ed. Paul H. Gobster and R. Bruce Hull (Washington, D.C.: Island Press, 2000). For constructive norms, see Eric Higgs, "What Is Good Ecological Restoration?" *Conservation Biology* 11 (1997), 338–48.

13. Bill McKibben, *The End of Nature* (New York: Doubleday, 1990).

14. Timothy B. Knopp and John D. Tyger, "A Study of Conflict in Recreational Land Use: Snowmobiling vs. Ski-Touring," *Journal of Leisure Research* 5 (1973): 6–17; and Stephen F. McCool and Justin Harris, *The Montana Trail Users Study* (Missoula: Institute for Tourism and Recreation Research, School of Forestry, University of Montana, 1994). For a larger survey and discussion, see John C. Adams, "Treadmarks on the Virgin Land: The Appropriate Role of Off-Road Vehicles in National Forests" (master's thesis, University of Montana, 1998).

15. Robert B. Reich, *The Work of Nations* (New York: Vintage Books, 1992); Al Gore, *Earth in the Balance* (New York: Plume Book, 1993); Bill Gates, *The Road Ahead* (New York: Viking, 1995).

16. Weinberg, *The First Three Minutes,* 154.

17. Robert Pack, *Before It Vanishes* (Boston: David R. Godine, 1989), 15.

18. Weinberg, *The First Three Minutes,* 154–55.

19. Gordon G. Brittan Jr. and David Healow, "Laughter and Fear in the 20th Century," paper presented at the Philosophy Forum, University of Montana, 17 November 1998, 5. See also John Morreall, *Taking Laughter Seriously* (Albany: State University of New York Press, 1983), 15–19.

2

Humans Supplant God, Everything Changes

Bill McKibben

When I worked on newspapers, we lived in constant fear of "burying the lead"—of sticking the real news in the sixth paragraph because we had been mesmerized by something flashier but less substantial.

I have thought of that idea a thousand times in the past decade, as I have watched politicos and academicians miss coming to grips with the story that really defines our time on this planet. Just in case you missed it, here is the lead that our descendants will stamp on the story of this moment: In the blink of an eye, and with hardly a thought, our species has come to the verge of dominating everything that happens on the surface of the planet.

When I say "in the blink of an eye," I mean it almost literally. Although the possibility of atomic war and then the insidious spread of chemicals like DDT gave us the first intuitions that our reach might be growing global, it is really with the advent of human-caused climate change that we have managed to run our fingers over every square inch of the planet's crust and every cubic meter of its atmosphere. And only for the past decade—fifteen years at the outside—have we been able to measure with real precision the changes we are causing, or to begin calculating their likely effects. In that time we have watched spring come an average of seven days earlier across the Northern Hemisphere; we have watched severe storms increase by a fifth across this continent;

we have watched the sea level start inexorably to rise as glaciers melt and warmer seas expand; we have watched animals begin to change migration and breeding patterns. We have watched, in other words, incredibly speedy and incredibly large changes in basic physical forces, all caused by the habits and appetites of one species. Ten years ago, the magnitude of these changes was still open to debate; even the notion of whether they were caused by human emissions remained a theory. Now, after a decade of frenetic research, only a tiny fringe of scientists remain unconvinced or unconcerned.

In precisely the same ten years—fifteen at the outside—we have watched genetic manipulation go from a small corner of the scientific enterprise to the very center of economic life. A decade ago, agronomists were conducting early trials of a few genetically engineered plants; this summer half the corn and soybeans planted in North America carry genes they could never have acquired with the kind of breeding human beings engaged in for the eight thousand years leading up to 1990. We now possess the power to redesign every biological system on the planet—every living thing, every chunk of creation—for our own convenience, pleasure, and profit. Maybe that's good news, maybe that's bad news, but it's definitely news.

The headline for the story about the last years of the Cretaceous period is pretty obvious: "Big Chunk of Rock Slams into Earth; Everything Changes." And the headline for our moment is equally stark. Forget the Internet, forget even the end of communism. "Humans Supplant God; Everything Changes." This is to our moment what civil rights was to my parents and the Second World War to their parents, and the Civil War to their great-great-grandparents—the overwhelming moral question of the moment, fraught with practical consequences, capable of yielding whole new ways of looking at the world.

So how come we have mostly missed it? How come this question is not preoccupying every philosopher and every planner and every theologian—not to mention English professor, policy maker, and educator? Why isn't it at the top of the agenda for every smart person who is theoretically concerned with the connection between human beings and something larger? Even the problems of social injustice that still plague us do not come with such pressing time limits. But if we fail to make—and quickly—major changes in the way we use fossil fuels, for example, then we might as well not even try. These problems have risen suddenly, and they must be solved suddenly, but instead they are left to lurk around the fringes of academic discourse.

Dolly, the world's first cloned adult mammal, developed by scientists at the Roslin Institute, Edinburgh, Scotland. Copyright Associated Press.

Partly we have failed to act because we have become pretty denatured. Those of us who live in cities and suburbs have been inside so long it is hard for us to notice that the outside is changing. The thinking professions have concentrated so relentlessly on the relations between human beings that the existence of something larger than us becomes a vague memory. The economy seems more real to us than the ecosphere.

And we have failed to act also because these problems are so big. We have learned to break issues down into ever-smaller pieces; grappling with fundamental threats to creation requires moving in the opposite direction. These threats demand big theories, grand commitments.

But even those academics who have noticed what is under way have managed to trivialize the moment we live in with a kind of clever logic that serves to mask our unprecedented situation. One group of environmental philosophers and historians, for instance, has repeatedly pointed out that human beings have always altered the world around them. It is hard to find an example of true "wilderness" on this continent, they say, because indigenous people periodically burned the land, changing its species composition somewhat. Therefore, what happens when you pave a wilderness, or when you turn it into a desert through global warming, is simply an extension of what the Amerindians did.

Or consider genetic engineering. We have bred corn for millennia, the argument goes, in an effort to give us plumper ears or hardier plants; therefore, snipping the genetic code off a beetle and inserting it in acorn plants is merely an extension of our old practice. No big deal. No need for a headline, business as usual, don't get excited.

It is perhaps too extreme to call this an intellectual con game, a scam. But not by much. Here is how it seems to work: A breakthrough is announced: Dolly the sheep has been cloned. In the papers, a series of "bioethicists" explain that it is really not all that different from what went before. The inventor says it is very difficult to do with sheep, and there would certainly be no reason to do it with humans. A presidential panel is appointed, never to be heard from again. A year later, the same technique works with mice, only this time the process is much easier; you can clone mice almost at will, which leads some other authorities to speculate that, in fact, human applications may not be so distant. But there is no need for worry: really, it is like what happened with Dolly. And so on, ad infinitum, each development seen merely as a small change from the one before. If you say, "Whoa, hold on, that seems like a big step to me, maybe we should talk about it," you are dismissed as backward looking.

This is all rather like the game where you need to transform, say, L-O-V-E to H-A-T-E by moving one letter at a time. Examined move by move, nothing very significant seems to be happening, but before long you have concocted something entirely new.

The point is that quantitative changes can be so large that they become qualitative. Yes, people have always changed the world around them; since we are slow and furless, we have had little choice. We have always altered the environments around our settlements, around our fields, around our hunting grounds; in recent centuries we have gone much farther. And yet there remains a visceral distance between those kinds of alterations and the new wholesale changes; witness, if nothing else, the orders-of-magnitude increase in the extinction rates we are now causing. Whole ecosystems, like coral reefs, are disappearing, and yet the frog-in-a-heating-pot school keeps telling us nothing much is new. Pay no attention to those bubbles, ignore that scalding sensation.

Or, better yet, celebrate it. Declare that in this grand new world we will all be cyborgs—and weren't we sort of cyborgs when we first put on eyeglasses!? And anyway, cyborgs are tremendous! And transgressive! And whatnot! In such fashion we have danced across this threshold, or nearly across it, with the unconcern of mariners crossing the dateline.

Not that all such developments, of course, are necessarily wrong. (You can even find those—although their numbers are steadily shrinking—who argue that global warming will benefit some regions by making certain kinds of agriculture easier.) My point is that we need to recognize the magnitude of the changes now under way. And if we are intellectually serious, and morally serious, we need to engage in far more soul-searching than we have done so far about whether they make sense or not. Soul-searching about whether we might want to confine genetic manipulation to certain problems of human health; whether we might want to respect the basic integrity of other species (which have always been defined precisely by who they will and will not share genetic material with); and whether we might want to make an all-out effort to lessen our impact on the earth's climate, even if it is too late to prevent it altogether.

It is odd how well the scientific process works—in a decade it has reached consensus on the question of whether climate change is occurring, and in the same decade it has forever uncracked the gene. By comparison, the other disciplines seem to travel with an almost oafish unconcern through these new waters, pretending that the old charts work just fine. Only on the fringes—among the nature writers, say—has the moment of our moment really been noticed. The irony, perhaps, is that our domination may carry the seeds of our own diminution. The forces we unleash by raising the temperature—and quite possibly the forces we unleash with what is essentially reckless genetic tinkering—may be so strong that they may overwhelm us. You cannot roll back winter forever before fundamental problems result. But when they do, there will doubtless be some new Ecclesiastes around to point out that there is nothing new under the sun.

2000

3

Too Much:
The Grand Canyon(s)
Lucy R. Lippard

There are too many Grand Canyons. There is the place itself and its stag-gering geography—the rims, the river in the Inner Gorge, the maze of side canyons, mesas, plateaus, forests, arroyos, vegetation, and wildlife, and all those hoodoos, columns, and *spires* (so-called by nineteenth-century devotees of the Church of the Wilderness). There is the no-nonsense (and topographically nonsensical) governmental gridding of ungriddable lands as the frontier fell away. There are the variously perceived canyons through which flow the never-ending ver-biage that attempts but never succeeds in seeing, let alone describing, this sight of sights. And at a deeper level, there are the interpreted canyons, the contested canyons. From these emerge our individual and collective psyches, reflected in the geographies of national history and personal experience. The abysses are epitomized by fundamentally di-vergent views of place and nature expressed by the canyon's Native peoples and by the ruling ethics of the National Park and Forest ser-vices, themselves often at loggerheads.[1] The Grand Canyon's macro-microcosmic multiplicity staggers retina and rhetoric.

"Without doubt our epoch . . . prefers the image to the thing, the copy to the original, the representation to the reality, appearance to being. . . . What is sacred for it is only illusion." Thus spake proto-postmodernist Ludwig Feuerbach, who died in 1872. Situationist Guy

Debord used this quote to kick off his classic *Society of the Spectacle*.[2] Today Debord's theories, mostly cleansed of Marxism, have recaptured attention, although they are rarely applied to landscape. Yet the Grand Canyon is a "natural" candidate for spectacle status, touted as one of "nature's" crowning achievements and therefore not to be blamed on civilization or the lack thereof. Debord taught us that our minds and eyes are trained by hegemony, Hollywood, et al., to treat spectacles as ahistorical objects. "The spectacle," he wrote, "as the present social organization of the paralysis of history and memory, of the abandonment of history built on the foundation of historical time, is the *false consciousness of time*."[3]

Picking up and running with this ball, postmodern academic criticism has taught us, ad nauseam, that the whole world is packaged for us, that nature is no exception, that we never see what is before us without an invisible frame courtesy of the mass media or even of great art and literature. We are, in other words, virtually forbidden to experience anything directly. Once we are persuaded by disaffected scholars to see all of nature as a sentimentalized theme park, a stage set, a backdrop, we are bereft of spiritual possibility, optimism, action. Debord predicted this misuse of his theory, warning that "undoubtedly the critical concept of *spectacle* can also be vulgarized into some kind of hollow formula of socio-political rhetoric to explain and abstractly denounce everything, and thus serve as a defense of the spectacular system. For an effective destruction of the society of the spectacle, what is needed is men [*sic*] putting a practical force into action."[4]

Actually entering the canyon, even in the constant company of twenty other people, as I did recently, can be an antidote both to illusion and cynicism; it can rip off the packaging and splinter the frame. (The ghosts remain: the package ghost—the experience we expect to have; the frame ghost—the pictures we expected to take.) But below, after the descent, down inside the canyons, the event is harder to cope with. Large and small vie for our attention; every rock, every cliff competes every minute with another, with "the view" ahead, the view behind, up and around, with the cacti and wildflowers, the despairing gestures of the red-blossoming ocotillo's skinny arms, with the harsh and melodious calls of ravens, canyon wrens, and gulls. Rough ground underfoot, we edge along chasms, wade into waterfalls, soak in grotto pools, clamber on slippery shelves, stumble through rocky streams and over parched boulders. And above and below it all, day and night, is the roar of rapids, burbling in the deep, surfacing in force, raising waves that can reach twelve feet at the infamous Lava and Crystal falls.

The real human construction is finally a sense of individual responsibility, although one component of any creative response is confrontation with previously imposed constructions. It is all too easy to drift lazily into fashionable preconceptions formed by others and find our fresh eyes filtered, our imaginations blocked. By allowing this to happen, we deprive ourselves of the kind of "authenticity" we think we are after. Yet as soon as we seek authenticity, we are nudged by education, sophistication, and modernity into anxiety about how much we are "influenced." Illusion is replaced by memory, equally susceptible to intellectual fashion.

Until I went past the rims, leaped over the edge, so to speak, the canyon did not interest me much beyond its obdurate identity as a spectacle, a melodramatic aesthetic, and an academic cliché. While I was on the river, I had not an "original" thought and barely a thought at all. I was lulled by being guided. I had no maps, although I love them. I didn't buy the long, thin fold-out with every rapid marked and intriguing tidbits of history and science in the margins. Somewhere in my mind, this had already been ordained as a mapless voyage. I had been asked to go along at the last minute and dropped everything to do it. For once, I hadn't done my homework. I didn't know where we were putting in or taking out. I didn't even know if it was a motorized or paddling or rowing trip. All I had time to think about was whether I had any nylon clothes that would dry quickly (only a leopard-skin pajama top that I never had the guts to wear) and whether my bent tent pole would hold up (it didn't). It was in many ways the ideal way to take the plunge. I was treading water.

The American West is full of stunning canyons, secreted among the flat grasslands and rolling high deserts, come upon with breathtaking suddenness at their brinks, glimpses into contrasting worlds of red rock, green trees, ruins, wildlife, and water. The Grand Canyon is layeredness taken past the equation, backdrops that flatten the imposing masses but never stop coming. It is "ageless" (not reducible to human scale, not easily comprehended) and constantly changing. "No matter how far you have wandered hitherto," wrote John Muir, "or how many famous gorges and valleys you have seen, this one, the Grand Canyon of the Colorado, will seem as novel to you, as unearthly in the color and grandeur and quantity of its architecture, as if you have found it after death, on some other star."[5]

Human construction of experience, fragmentation into spectacle, is a weapon, armor, or refuge from the terrible possibilities of chaos, of unconstructed, unimagined experience. In retrospect, the gap between general and specific is what catches my imagination. It seems to offer a bridge between theory and practice. Action is an antidote to the indescribable. *"I find that I tell more about what we did than what we saw,"* wrote Marion Smith, after a Grand Canyon raft trip with river woman Georgie White Clark. "To describe the scenery is quite beyond me and almost, I believe, beyond the camera. The distances up above, way ahead, and far behind, are so vast, the great canyon so alive somehow, that its true scale and impact can only be approximated even with motion pictures. One must go there to know it."[6] The relations between doing and seeing, action and vision, construction and perception, lie at the core of the Grand Canyon experience—or experiences.

In cultural terms, landscapes only come alive, in fact only become *landscapes,* when they are focused upon, when they become specific, when humans begin to interpret them, like the hoary tree falling in the forest. A landscape experienced as generalization, the frequent fate of the Grand Canyon, is not itself. As Paul Shepheard writes, ". . . although everywhere the world is the same as itself, landscape is nowhere the same as itself: you have to show landscape by example, because as a subject it won't reduce to fundamentals; it won't *trivialize.* . . . Geography is global. Chorography is regional. Topography is local."[7] You can lose your way in the specific because there are too many paths, too many featureless thickets, too many options, too many box canyons and shelves with no exit. You can also lose your way in the general, because there is no path.

The Grand Canyon as human construction, or cliché, is familiar to everyone. Say the magic words and a single image appears in the mind's eye—the view from the North Rim. It is an image formed by images for well over a century, an image that has survived fluctuating conceptions of nature and our responsibilities to it. Before the invention of halftones, the public saw the canyon through the eyes of engravers taking liberties with photographic originals, exaggerating heights and grades. Then came color—the railroad propaganda and the calendars, red purple brown gold orange and blue cliffs layered by light and mists into infinity—beautiful images, and boring in their beauty.

It has been argued that there is no such thing as wilderness. It has been argued, equally viably, that there is. However our notions of wilderness may vary, and have varied culturally over the recordable centuries, I think most of us can judge when we are within it, or within our own notion of what it should be, or approaching that state of mind. The level of uneasiness increases, matched by the level of exhilaration. Wilderness, even as a generalization, is not a place you can look at or even into. Standing on a rim, you may say you have been *at* the Grand Canyon, you may even say you have *seen* the Grand Canyon, but without the descent (and all it implies), you have not been *in* the Grand Canyon. Opening up the Wilderness, they called it in frontier days, sexual innuendoes intact—as in virgin land, rape of the land.

William Cronon says that in the eighteenth century there was a "sense of wilderness as a landscape where the supernatural lay just beneath the surface. . . . God was on the mountaintop, in the chasm, in the waterfall, in the thundercloud, in the rainbow, in the sunset. One has only to think of the sites that Americans chose for their first national parks—Yellowstone, Yosemite, Grand Canyon, Rainier, Zion—to realize that virtually all of them fit one or more of these categories."[8] (God did not visit the swamps and the grasslands until decades later.) In the nineteenth century, however, Americans found God wanting in a few details and took over the job for themselves. In the twentieth century we lived out the techno fantasies and have lived to regret some of them. At millennium's end, the Grand Canyon

Lava Falls, the largest rapids in the Grand Canyon. Copyright Tim Thompson/CORBIS.

is as often as not valued for its "spiritual" gifts. We have come almost full circle.

There has always been internal manipulation of landscape, and there has always been external aesthetic control. The excesses to which contemporary agonizing over intellectual "constructions" has gone are examined by Jon Margolis, writing in *High Country News*.[9] He cites attacks on wilderness legislation not only from the political right but also from the left: "Well, actually, they're from the post-modernists, which is not the same thing. It's not that these critics are against wilderness, exactly; they're just disturbed by the idea of wilderness." Then, with grudging approval, he quotes Cronon saying that the real problem lies not in wilderness per se but in "what we ourselves mean when we use that label." After making wholly justi-fied fun of some incomprehensible academic jargon on the subject, Margolis finally concedes that "having to respond to a more complex critique has forced pro-wilderness troops to sharpen their scientific, cultural and political case." Ideally, these critiques would themselves be couched in collaboration with conservationists, lending themselves to a clearer "reading" of landscapes like the Grand Canyon.

An understanding of the society of the spectacle is useful, and a clear sense of how we are manipulated by representation is a neces-sary tool for surviving postmodern life, pointing up aspects of the contemporary experience even as we deny that such a thing can be unmediated. But knowledge to the extreme can also be destructive. Exaggerated skepticism (borderline cynicism) can leach away our in-nocence, smooth off the rough edges that allow our gears to fit into those of wild places, that allow us to understand our familiarity with what we persist in calling nature. When we forget to include our-selves, we are cut off from our surroundings and our cohabitants. No place lends itself more easily to these affectations than the Grand Canyon(s), especially the one conceived from the rims. Here are places that are "inhuman"—godlike or demonic. We cannot reason with such places, our culture seems to imply, so we had better try to overpower them. Vulnerability is a thorn in our flesh. Every tale of flash floods, falls from crumbling cliffs, lightning strikes, drown-ings, heatstrokes, summer hailstorms, unbearable heat and cold has its moral and calls out for Control.

Damming great rivers is an unsubtle way to reconstruct their mean-ings. The watering of the arid West—a remarkable and often repellent story brilliantly told by Donald Worster *(Rivers of Empire)* and Marc

Reisner *(Cadillac Desert)*—includes the damming of Glen Canyon and the irony of naming the resultant Lake Powell (a.k.a. Lake Foul) after the man who warned that the waters of the arid West should remain in the hands of locals who understood it. And as if Lake Powell were not enough, as if the canyon had to be punished for still thumbing its nose at those who would dominate it, plans continued to be made to dam the Grand Canyon itself. It was, in fact, saved by the tragedy of Lake Powell and by the deflection of a dam at Echo Park in Dinosaur National Park. The proposed dams were stopped by a public shift in attitude away from water exploitation and toward wilderness protection.[10] This happened in the 1950s, when the canyon underwent "an immense cultural metamorphosis," becoming again, though the imperial overtones this time were somewhat muted, part of a national "geography of hope."[11]

When conservationist David Brower said, "If we can't save the Grand Canyon, what the hell can we save?" he acknowledged the necessity of publicity, a ploy that often backfires.[12] Singer-songwriter Katie Lee noted that "the more people you get to fight for the rivers, the more people you have to take to the rivers, thereby ruining them in order to save them from destruction."[13] The millions of tourists who visit the Grand Canyon each year, myself included, go there for a relatively safe experience of risk, a temporary sense of freedom within the bounds of reason and security. These days the "spiritual" is also much in demand. And who can escape it in the Grand Canyon? The place produces stories not merely of action as the macho adventure magazines would have it, but also of interaction.

With a certain logic of contrast (and incomprehensible overload), tourists often take a package tour combining two disparate spectacles: the Grand Canyon and Las Vegas—gambles of a different nature. "Capitalist production," wrote Debord, "has unified space, which is no longer bounded by external societies. This unification is at the same time an extensive and intensive process of *banalization*. . . . A byproduct of the circulation of commodities, tourism, human circulation considered as consumption, is basically reduced to the leisure of going to see what has become banal. . . . The economic organization of the frequentation of different places is already in itself the guarantee of their *equivalence*. The same modernization which has removed time from travel has also removed it from the reality of space."[14]

Like tourists, contemporary writers must be wary of the "superficial mysticism, sentimentalism, and loss of critical faculties," which are an occupational hazard when we are connecting with something larger than ourselves. Is Thoreau's "vital and intense sense of connection to nature" possible in a skeptical postmodern era?[15] Anyone who can avoid a certain romanticism in the Grand Canyon is probably hopeless. John Wesley Powell, a remarkably tough guy, wrote, "A mountain covered by pure snow 10,000 feet high has but little more effect on the imagination than a mountain of snow 1,000 feet high—it is but more of the same thing; but a facade of seven systems of rock has its sublimity multiplied sevenfold."[16] Are we no longer allowed such responses because we have learned their transparency? Can't we also maintain their transcendency? The hyperbole resorted to by Muir, Powell, and endless others is normal. The place itself is off the charts. It resists casual treatment even when reduced to scientific "truths." We know that the boring and astounding calendar picture is not just a geology textbook. The Grand Canyon's all-consuming scale leaves little energy for facile consumption. Writers like to mull over palimpsests, but enough is enough. The extravagantly layered strata flaunt their dangerous illegibility, so laden with efforts at legibility—with artifice and generalization—over the past century that they threaten to take up all the space our mythology has allotted to the Grand Canyon. They interfere with the firsthand experiences we value—so-called authenticity.

Our contemporary models, although many, offer little help in coping with the (artificial) divide between subjective awe and objective reining in of emotion. Finally, it is the "Thoreau of the West," Edward Abbey—macho, bigoted, intolerant, arrogant, and irritating as he often was—who most convincingly found words for his passion for the canyonlands, who found a language that combined poetry and outrage. Sometimes I hate to like Abbey's writing so much. While he can be as sentimental and rhetorical as the next applicant, he was rarely "soft-headed," and frequently, almost involuntarily, lyrical. In his essay "Down the River with Henry Thoreau," he insisted that the Sage of Concord "becomes more significant with each passing decade. The deeper our United States sinks into industrialism, urbanism, militarism—with the rest of the world doing its best to emulate America—the more poignant, strong and appealing becomes Thoreau's demand for the right of every man, every woman, every child, every dog, every tree, every snail darter, every lousewort, every living thing, to live its own life in its own way at its own pace in its own square mile of home. Or in its own stretch of river. . . . The village crank becomes a world figure. . . . Truth threatens power, now and always."[17] At heart, however "postmodern" (which is a way of coping with modernity as much as it is a theory), however distrustful of words like "truth" we may have become, I believe that most of us feel as Abbey does, although few go on to give it time or energy.

Before techno bliss, simulated landscapes were created as photo backdrops, dioramas, and historical "machines," like Thomas Moran's late and very large paintings of the Grand Canyon, shown as autonomous "exhibits" rather than as parts of an exhibition. "Topography in art is valueless," said the artist.[18] "The Canyon's rim was American art's greatest gallery and its greatest pulpit," writes Stephen Pyne, author of the crisply brilliant compendium *How the Canyon Became Grand*. There have been heroic attempts to depict the canyon by painters and photographers (not sculptors, because it is being done already, too well, right there before our very eyes). They are all men—Moran, H. B. Mollhausen, John Hillers, Timothy O'Sullivan, William Henry Holmes, Carl Oscar Borg, Gunnar Widforss, and finally Eliot Porter, who tried to resacralize the place in the 1960s in the face of more dam proposals. Eventually, progressive artists gave up on the Grand Canyon, which had become a "celebrity . . . a museum piece . . . culturally moribund," and therefore vulnerable to exploitation.[19] Despite the attempts of a few bold artists (such as Susan Shatter) to represent the canyon, the dry realism now fashionable is as inadequate to the task as Modernism itself has proved. Pyne blames the triumph of commercial art on Modernism's disdain for its audience: "High culture had to compete with the purveyors of the Canyon as experience, not simply as idea. . . ."[20]

Inherent, if invisible, in the Grand Canyon's apocalyptic image is the consciousness of destruction that Barbara Novak identifies as a mark of nineteenth-century nature painting. There is a remarkable similarity in the content, if not the style, of aesthetic alarm over the past century. "The axe of civilization is busy with our old forests. . . . What were once the wild and picturesque haunts of the Red Man, and where the wild deer roamed in freedom, are becoming the abodes of commerce and the seats of manufactures . . ." wrote J. F. Cropsey in 1847.[21] Today, the threatening axe of previous centuries has become the still more menacing bulldozer and backhoe, the intruding road has become the helicopter, the specters are now pollution, crowds, and loss of habitat. Chuck Forsman, in his *Arrested Rivers* series of photorealist paintings that depict the dammed (and damned) rivers of the West, belies the placid, complacent views expected of conventional landscape painting. Janet Culbertson's poignant canvases of stark, barren landscapes, inhabited only by billboards that are themselves paintings of lost idyllic scenery, give nostalgia pause. Could anyone in the nineteenth century have imagined how far things would go? Can we imagine the future pictured by Culbertson, where only the image remains?

It has been suggested that designers may be tempted by the "increasing power to control and modify nature."[22] What kind of unprecedented work might they come up with? It gives me the creeps to think about the extreme limits of this proposition. A landscape designer stands on the North Rim, looks outward, downward, like a nineteenth-century landscape painter with brush poised between palette and canvas. Let's see, a bit more red wall there . . . take out that awkward parapet here, that boring schist. . . . There now. *That's more like it.*

Like what? Like the nineteenth-century painter's rendering of the sublime? Like the mid-twentieth-century photographer's "reality"? Like the conceptual artist's ironies? Like the publicist's seduction and hyperbole? Into what chink in those towering, spotlit walls can we fit our own "views"? What next? A plastic Grand Canyon theme park to replace the one dammed and devastated within the church of progress?

The graffiti is on the red wall, so to speak. A recent press release from a New York gallery describes the work of Los Angeles artist Jacci Den Hartog, which "explores the potential of manufactured materials to simulate nature and reconstitute a comparable sublime." Like the existing Japanese indoor beaches and ski slopes, this makes the mere *idea* of the human construction of nature look tame. Not quite virtual reality (that's next) is contemporary artists' response to postmodern constructionism. Andrea Zittel's *Point of Interest*—two giant faux rock outcroppings (concrete over steel)—is the latest and largest in a line of art rocks over the past thirty years or so. (Earlier simulators include artists Reeva Potoff and Grace Wapner, as well as the designers of zoos and pet rocks.) Zittel's sculpture is, according to the Public Art Fund's 1999 press release, "a reminder that Central Park itself is a meticulously planned natural environment built for the enjoyment of city-dwellers . . . a playful critique of late-20th-century society's 'action adventure' uses of nature (from extreme mountain climbing to the increasingly popular 'Eco-Challenge'). . . .

Point of Interest serves as a reminder that our perceptions of nature are constantly being reinvented and often reflect the values and ideas of society itself."

I imagine an exhibition of the millions of amateur snapshots taken in the Grand Canyon. The viewer would walk miles through corridors of canyon impressions, edge to edge, all more or less alike despite widely varying skills and technologies. My own photographs were lousy (artist friends didn't do all that much better). There was a tremendous disjunction between what was seen and what was represented. This is probably endemic to all photographers of the Grand Canyon, even the almost first and perhaps greatest of them, the methodical genius Timothy O'Sullivan. I managed not to photograph most of my favorite places, a wise, if inadvertent, decision dictated by the fact that my old manual Olympus was wet a lot of the time and the shutter stuck at random. In any case, visual experience is not always about making pictures—in the mind, the canvas, the camera. Pictures were only the end product of an experience that was as kinesthetic as it was visual.

Nor did I make any attempt to remember those 280 miles of amazement. I didn't "journal" (for a journalist, that's a buswoman's

Picnic on the edge of the rim, Grand Canyon, 12 February 1983. Photograph by Mark Klett.

holiday . . . and it's not a verb anyway). If I let my mind go blank, I can recall the general and the specific, but not the connections, not the whole. Images resurface—peaceful winding expanses of water more green than blue; the ever-changing rock show and its reflections in the emerald, jade, turquoise, slate, salmon to brown water; the extraordinary panoramas; the wild misty turbulence, the wind, the spray, the permanently wet butt, the high waves and fleeting exhilaration of the rapids, almost dreamlike in their brevity and intensity.

Only when I returned home and started reading Powell, Abbey, the expedition reports, the women's stories, the historians' stories, did the constructions loom. The canyon became realer to me in some ways than when I was there. Now, I realized with a certain sadness, innocent perception was beyond me. Had I read all this to begin with, I would have been walking through a much solider looking glass. But ignorance suited me. What could I have known about this place before having been inside it? It is not for nothing that the river is a preeminent symbol of life. But the narrative flow built into this trip still eludes me. There wasn't time to consider all that space. We hurried. We had to get on down the river.

2000

Notes

1. For a thorough examination of the conflicting roles of government agencies in the canyon, see Barbara J. Morehouse, *A Place Called Grand Canyon: Contested Geographies* (Tucson: University of Arizona Press, 1996).

2. Guy Debord, *Society of the Spectacle* (Detroit: Black and Red, 1970).

3. Ibid., 203.

4. Ibid., 158.

5. Stephen J. Pyne, *How the Canyon Became Grand: A Short History* (New York: Penguin, 1998), 70.

6. Richard Westwood, *Woman of the River: Georgie White Clark, White Water Pioneer* (Logan: Utah State University Press, 1997), 156, my italics.

7. Paul Shepheard, *The Cultivated Wilderness* (Cambridge, Mass.: MIT Press and Graham Foundation, 1997), iv.

8. William Cronon, "The Trouble with Wilderness, or, Getting Back to the Wrong Nature," in *Uncommon Ground* (New York: W. W. Norton, 1996), 73.

9. Jon Margolis, "Do You Want More Wilderness? Good Luck," *High Country News,* September 27, 1999, 5.

10. James L. Wescoat Jr., "Challenging the Desert," in *The Making of the American Landscape,* ed. Michael P. Conzen (New York: Routledge, 1994), 202.

11. Pyne, *How the Canyon Became Grand,* 151.

12. Quoted in ibid., 153.

13. In Betty Leavengood, *Grand Canyon Women: Lives Shaped by Landscape* (Boulder, Colo.: Pruett Publishing Co., 1999), 195.

14. Debord, *Society of the Spectacle,* 165, 168.

15. This question and the prior quotation are extracted from a publication brief written by the editor of *Harvard Design Magazine* for the issue of the magazine called "What Is Nature Now?"

16. John Wesley Powell, *The Exploration of the Colorado River and Its Canyons* (New York: Penguin Books, 1997 [1875]), 380.

17. Edward Abbey, *Down the River* (New York: Penguin Books, 1991), 36–37.

18. Quoted in Pyne, *How the Canyon Became Grand,* 89.

19. Pyne, *How the Canyon Became Grand,* 136.

20. Ibid., 117.

21. Quoted in Barbara Novak, *Nature and Culture* (New York: Oxford University Press, 1980), 5.

22. This suggestion was made by the editor of *Harvard Design Magazine* in the guidelines for the issue of the magazine called "What Is Nature Now?"

4

What Do We Make of Nature Now?

Catherine Howett

In August Strindberg's 1902 *Dream Play,* a hovering angelic presence observing the action on stage repeats the refrain "human beings are indeed to be pitied"—a mantra peculiarly appropriate to the mood of thoughtful men and women almost a century later. The seductive euphoria of American prosperity and unchallenged global power has prompted unsettling comparisons with both the Gilded Age and the Roaring Twenties—historic moments during which the rich and comfortable indulged their pleasures, largely indifferent to any risk of catastrophic social, economic, or political upheaval. Today, of course, we must add environmental disaster to the list of nightmares hovering offstage, like Death and Plague in a medieval morality play. Call it *Dame Nature's Revenge.*

Perhaps what most troubles those who feel anxious about the future is a sense of helplessness about our capacity to protect ourselves or the people and places we care about, much less the human race and the planet. It is not simply the greedy or fanatic we have to fear, after all, when well-intentioned bioengineers at a lab in some distant Happy Valley are tinkering with the food we eat and a lot fewer frogs are sounding off on summer nights at the lake.

The complexity and scale of contemporary environmental problems have provoked Wendell Berry, whose quietly passionate voice has

Asher B. Durand, *Kindred Spirits,* 1849. Oil on canvas. Courtesy Crystal Bridges Museum of American Art, Bentonville, Arkansas.

drawn many to the cause of more enlightened stewardship of the earth, to declare recently that he has had it with high-minded movements. His disaffection springs from bitter awareness of how often such groups become self-righteous, freighted with bureaucracy, willing to compromise critical values, or else simply fail to achieve the desired ends:

> The proper business of a human economy is to make one whole
> thing of ourselves and this world. To make ourselves into a practi-
> cal wholeness with the land under our feet is maybe not altogether

possible—how would we know?—but, as a goal, it at least carries us beyond hubris, beyond the utterly groundless assumption that we can subdivide our present great failure into a thousand separate problems that can be fixed by a thousand task forces of academic and bureaucratic specialists.[1]

Berry tries to temper the impact of his pessimism by exhorting the committed to work toward nothing less than a restructuring of the world economy through a movement that is less a *movement* than an aggregation of individual acts of bearing witness, of showing, in daily life, "respect for this earth and all the good, useful, and beautiful things that come from it." He suggests that by practicing "domestic arts . . . stationed all the way from the farm to the prepared dinner, from the forest to the dinner table, from stewardship of the land to hospitality to friends and strangers," a grassroots revolution might be set in motion. At the least, those who pursue this course will enrich their own lives.[2]

It is easy to understand the appeal of such a proposal—to align oneself with an invisible army of saints pledged to saving the world through modest acts of virtue, and thus to spare oneself association with organizations that inevitably will go awry in ways large or small, since any group will include a few who are cowardly, venal, stubborn, foolish, or worse. But is this proposition not another manifestation of hubris, in its confident presumption that personal acts and judgments will be purer and more effective than the mucky struggles of those willing to put up with endless meetings and conferences, marches and protests, campaigns and letter writing, money and time poured out—and often, admittedly, with little to show for it?

The American environmental movement had its philosophical grounding in nineteenth-century New England transcendentalism and particularly in the idea of the natural world as the primary source of enlightenment, morality, and contentment. Now we have Berry's growing suspicion of collective action and his embrace of what sounds like Emersonian "Self-Reliance"—not prideful or selfish egotism but faith in individual belief and action, which can draw on the universal wisdom within each soul. Moreover, for Emerson as for Berry, a "whole" person does not view a career as the way to earn a living but as a form of service to humanity. Thoreau pressed still further the need for independent decision making and actions, confident that any soul "true . . . unto itself alone, and false to none" needs nothing from the body politic, owes nothing to the authority of laws or social organizations of any kind.

The transcendentalists' romantic and utopian vision of primal nature and human nature remains viable in American culture, and not just among environmentalists. Its potency—it is a simple idea, promising sweet liberation and suffused with a glow both rational and spiritual—informs our mythic life as a nation, finding expression in countless images, songs, and stories. This vision is also manifest in a persistent emotional bias against cities, because they usurp the place of nature and demand physical closeness and interdependency rather than self-sufficiency. In *Walden,* Thoreau argued that "in nature, not in the town, the individual may walk with the Builder of the Universe." In the same spirit, in his funeral oration for the painter Thomas Cole, William Cullen Bryant admonished America's artists to preserve the vestiges of sacred wilderness from "the axe of civilization," from "the abodes of commerce and the seats of manufactures."[3] In the next century, Frank Lloyd Wright, his sensibility shaped by the transcendentalists and Ruskin, urged his countrymen to move their families out of cities and to live independent lives close to nature, as far from other people as they chose, and to dwell there in "natural" houses. Wright countered European Modernism's embrace of the machine as emblem of the new age of technological revolution with an organic metaphor—the building growing out of its site as a tree grows, penetrated by light and air, opening out to the surrounding landscape.

The myth that celebrates retreat from a world dominated by human culture to a more pristine natural environment has had strong appeal for certain artists. Two of America's early Modernist painters, Arthur Dove and Georgia O'Keeffe, felt compelled to quit city life and to live closer to nature. Dove believed that the curiosity and joy he felt in contemplation of the natural world as a boy growing up on a farm in Geneva, New York, was the abiding wellspring of his art. Years later, he recalled an epiphanic experience that occurred when, after returning homesick from a painting tour in the French countryside, he had gone camping in the Geneva woods. He described the rapture of "waking up looking in the woods for motifs, studying butterflies beetles flowers," and suddenly knowing not only that he must abandon impressionism, but also that he would work toward a style truly his own through a deeper seeing better suited to abstraction.[4] When he returned in the 1930s to live on the farm, the seclusion, simplicity, and rhythms of a life that filled his eye and mind with images of the natural world nurtured his most productive period, despite severe physical hardships and poverty.

His friend and fellow artist within the Alfred Stieglitz circle, Georgia O'Keeffe, felt a great sympathy between her own art and that of Dove, whom she described as the only American painter "of the earth."[5] Of the New Mexico home—her Walden—to which she eventually retreated, leaving behind the security and comfort of eastern urban culture, O'Keeffe said: "Sometimes I think I'm half mad with love for this place. . . . My center does not come from my mind—it feels in me like a plot of warm moist well-tilled earth with the sun shining hot on it."[6] She took pride in the fact that every inch of the house in Abiquiu had been smoothed by her own hand, and she made a private world for herself amid an extraordinary landscape, which became the generative locus of her life and work.

The far-reaching social and political revolutions of the 1960s re-claimed the myth of salvation through nature for a new generation. Nice distinctions among those who "dropped out" and "turned on" have tended to blur in popular memory, although the willingness to challenge assumptions about God, country, and middle-class mores while privileging one's intuitive sense of right and wrong was common to very different sorts of people: conscientious objectors leav-ing home for Canada, ordinary people of exceptional courage riding buses toward a beating and jail, WASP youngsters fleeing suburbia to fast and chant with the Hare Krishna, stay-at-home moms heading back to school or a job, as much as Haight-Ashbury hippies trip-ping on LSD. Dressed as monks, pioneers, or peasants, bearded or braided, a fair number also went "back to the land," actually or just in imaginative longing. However historians or the broader society may choose to judge those tumultuous years, no one can doubt that the cumulative effect of so many movements and actions both indi-vidual and collective profoundly changed this country.

Two recent books offer marginal notes to that history germane to this discussion. One is a biography of Rosa Parks, the African American who became famous for her refusal to give up her seat to a white person on a segregated bus in Montgomery, Alabama, in 1955.[7] Its author disputes the conventional view of Parks as an ac-cidental activist—an ordinary woman who one day, weary and hu-miliated, summoned the courage to say "no," an act of defiance that resulted in the Montgomery bus boycott. Such a gesture, were it the whole story, would certainly be one of those individual acts of witnessing that Berry sees as exemplary. The fact is, however, that Parks had for a dozen years before 1955 been active in the local NAACP chapter; had learned of earlier organized challenges to segregation, including a bus boycott in Baton Rouge, Louisiana, two years before her own arrest; and had attended a ten-day training course for civil rights activists at the Highlander Center in Tennessee the preceding summer. Her biographer argues that the romantic view of Parks as "the mother of the civil rights movement" distorts a more important reality, namely, that she had enlisted years before in the movement for change, and that her "tremendously consequential act might never have taken place without all the humble and frustrating work that she and others did earlier."[8]

The second book is Jedediah Purdy's *For Common Things: Irony, Trust, and Commitment in America Today,* one of several recent works by twenty-somethings attacking what they perceive as the pervasive cynicism of contemporary American society.[9] This fashionable ironic posture, according to Purdy and others, corrodes any possibility of trying to change the world for the better by mocking the efforts of those who espouse such old-fashioned virtues as hopefulness and sincerity. The issue of literary quality notwithstanding—the book has been criticized as naive, simple-minded, overwritten—Purdy (named for a mountain man of the American West) can be seen as an embodiment of the heroic and innocent child of nature come to manhood as the conscience of his culture precisely because he has remained isolated from it. Purdy's parents have been described as former hippies who, influenced by Berry's ideas, moved to a West Virginia farm "to revive an agrarian ideal, to turn their backs on the hollowness of mainstream living . . . to grow their own food in the wilderness and make sense out of life with both hands."[10] They lived without electricity until the late 1970s and without indoor plumbing until 1989, mostly homeschooling their children until Jedediah left for Exeter, then Harvard, then law school at Yale. An avid student of Thoreau, Purdy admits that his passion for the natural world reflects "animist sensibilities." His mother, however, has recalled with some amusement that the son who sees nothing but goodness in nature "was born judging people. . . . It was like having a priest in the house."[11]

A comparison of Parks and Purdy is not intended here, although a young man often described (without irony) as a genius might eventually influence the intellectual discourse of his time and even the cultural changes that inevitably follow such discourse. Parks's commitment to the civil rights movement began shortly before the 1960s; the Purdys' decision to withdraw into the "wilderness" came just after. Both were precipitated by forces that crystallized in the great debates of that decade.

Not surprisingly, those forces sparked important new directions in the art world in the same period. Most significant for the resurgence of romantic attitudes toward nature was the movement to which the tag "earth art" was attached early on, recalling the importance to Georgia O'Keeffe of earth as a metaphor for both fecundity and groundedness. Painters like Robert Rauschenberg and Jasper Johns had already erased the lines separating the art object from real space. Other artists began to explore ways to overcome visual and spatial conventions that posited a work of sculpture as a static and self-contained "object," rather than as what Robert Morris described as a "landscape" expressive of process rather than product. And because museums and galleries seemed complicit in turning sculpture into a conveniently fungible three-dimensional commodity, some artists were inspired once again to seek renewal in nature, moving outdoors, heading west metaphorically and in some instances actually. Michael Heizer, for example, claimed that by working at remote sites in the desert he found "that kind of unraped, peaceful, religious space artists have always tried to put in their work."[12] The critic Philip Leider described the experience of a group of friends who visited Heizer's *Double Negative*, a work measuring 1,600 by 50 by 30 feet that had required the excavation of 240,000 tons of rock and sand:

> [It] was on a giant mesa behind the town of Overton, Nevada. We were all expecting something strong, but none of us were prepared for it, as it turned out. . . . The sun was down; we wound up slipping and sliding inside the piece in the dark. [It] was huge, but its scale was not. It took its place in nature in the most quiet and unassuming manner, the quiet participation of a man-made shape in a particular configuration of valley, ravine, mesa, and sky. From it, one oriented oneself to the rest in a special way, not in the way one might from the top of the mesa or the bottom of the ravine, but not in a way competing with them either. The piece was a new place in nature.[13]

The description of gentle accommodation with nature is surprising for a heroically scaled work whose entire configuration can be seen only from the air. It suggests that this enormous linear excision in the earth—its two channels facing across the chasm like conduits of an invisible force leaping the void—prompts an aesthetic experience akin to those encountered in certain natural landscapes. The eighteenth-century English philosopher Edmund Burke is usually credited with developing the distinction between *beautiful* and *sublime* landscapes, the latter possessing a scale or grandeur so overwhelming as to inspire awe, rapture, or terror; later literati of the landscape gardening school added the category of the *picturesque*.

Nineteenth-century Americans with even a little interest in literature and art were familiar with these aesthetic categories, through reading and through the example of the Hudson River school of landscape painters—Thomas Cole, Asher B. Durand, and such disciples as Frederic Church—who substituted American subjects for the beautiful of Claude Lorrain, the sublime of Salvator Rosa. But it was the horticulturist, author, and editor Andrew Jackson Downing who made the canon accessible to a broad middle-class audience of homeowners, by insisting that the design of the domestic landscape, whether grand or modest, should express one or the other of these landscape types. Since the sublime could hardly be domesticated, Downing advised "gentlemen of taste" to choose between the beautiful and the picturesque, and to base the decision on the natural character of the site. The stakes in ignoring the admonishments of the esteemed Mr. Downing were made higher by the fact that such matters of taste had by the mid-nineteenth century, thanks to the transcendentalists and later to Ruskin, taken on moral weight—a choice between right and wrong with the well-being of one's family and even the republic itself at risk. Could a man in touch with the indwelling divine be insensitive to how his domestic landscape might instill virtue and a love of beauty in his children, while contributing as well to the beauty and refinement of his country?

In a curious way, and despite his fawning Anglophilia and self-conscious gentility, Downing—a wheelwright's son who married the great-niece of John Quincy Adams and rose to national prominence living by his wits—respected the common man, took pride in liberal democracy, and committed himself generously to such causes as the movement for public parks. In this he anticipated Frederick Law Olmsted and other social reformers who labored to move the "Great Experiment" toward what Olmsted described in 1853 as "a form of government in which . . . there are no privileged orders; no ruling class; in which the laboring class is being made, equally with the capitalist . . . the recipient of governmental power."[14] The Civil War would test this interpretation of American democracy against the Jeffersonian model—that government is best that governs least—to which the South had eagerly turned. Romantic individualism, centered in the human encounter with godly nature within the self and yet transcendent, was strongly identified, as we have seen, with an intense privateness, with the isolation of superior individuals, the poets and high priests of exquisite sensibility. So it took remarkable insight for Olmsted and other liberal and progressive reformers of the postwar decades first to insist that government must be empowered to guide an increasingly complex urban and industrial society toward a genuinely democratic civilization, and then to work energetically toward that goal.

More than a century later, the most articulate spokesman for early efforts to escape dependency on the systems of patronage that control the production, marketing, exhibition, and sale of art, and to work instead in and on landscapes, making "new places in nature," was Robert Smithson, best known for his 1970 *Spiral Jetty,* on the north shore of the Great Salt Lake in Utah. Smithson challenged the perception of art as an elitist luxury beyond the means and understanding of ordinary people; art should be, he believed, fundamentally social and democratic. To achieve such an art—authentically grounded in the past and present life of a place and a community—the foolish "spiritualism" (Smithson's term) that refuses to acknowledge the human place in nature had to be overcome: "The farmer's, miner's, or artist's treatment of the land depends on how aware he is of himself as nature. . . . When one looks at the Indian cliff dwelling in Mesa Verde, one cannot separate art from nature."[15]

The article from which this quotation was taken appeared shortly before Smithson's untimely death in the summer of 1973; it was the last of the provocative essays in which the artist articulated the conceptual bases of his art and that of the movement for which he had been a driving force and "powerful conscience."[16] Smithson discovered in Olmsted—whom he described as "America's first earthwork artist"—an exemplar of his own attitude toward nature and a model for his goal of working with government and industry to achieve "a democratic dialectic between the sylvan and the industrial." Moreover, his reading led him to those eighteenth-century authors whom Olmsted had claimed as his own "professional touchstones." The writings of such contributors to the picturesque debates as Uvedale Price, William Gilpin, and Richard Payne Knight had helped Olmsted to recognize that the natural world should not be represented as static, idealized, or sentimentalized. Rather, the original notion of the picturesque, Smithson learned through Olmsted, accommodated a vision of nature as an arena of chance and change, of dynamic transformations of energy and matter at scales ranging from the infinitesimal to the vast, essentially unknowable at both extremes.

Against Olmsted's appreciation of the complexity of natural systems, including those processes in which humans participate, Smithson posed the one-sided idealism of those who appropriated a standard of ecological concern to save the environment from art as well as from industry. The eighteenth-century theorists, Smithson believed, sensed the need for a "concrete dialectic between nature and people." Photographs of the site of New York City's Central Park before implementation of the "Greensward" plan of Olmsted and Calvert Vaux reminded him of strip mines in southeastern Ohio, and he compared Olmsted's undertaking, which required moving some ten million cartloads of earth, to his own interest in recycling landscapes:

> The site of Central Park was the result of "urban blight"—trees
> were cut down by the early settlers without any thought of the future.
> Such a site could be reclaimed by earth-moving without fear of up-
> setting ecology. . . . The best sites for "earth art" are sites that have
> been disrupted by industry, reckless urbanization, or nature's own
> devastation. For instance, the Spiral Jetty is built in a dead sea, and the
> Broken Circle and Spiral Hill in a working sand quarry. Such land
> is cultivated or recycled as art. On the other hand, when Olmsted
> visited Yosemite it existed as a "wilderness." There is no point in
> recycling wilderness the way Central Park was recycled.[17]

It is safe to say that few people in the environmental design professions—few architects, even fewer landscape architects—were reading *ArtForum* in those years; thus, the import of Smithson's death at thirty-five, when he was grappling philosophically and artistically with questions of how human making relates to nature, was not appreciated by those to whom, whether they knew it or not, it mattered most. It mattered not just because Smithson was "digging through the histories," as he described it, searching out the sources of how we came to think about nature as we do, examining alternative conceptions that might help us to think more perspicaciously about the relationship between human culture and the rest of nature. His death mattered because he took sharp aim at the romantic myth that sees nature as ineffably grand, good, and godly, best encountered alone and in quiet out in the wilderness or at least out in the country. Debased popular expressions of this myth require us to ignore those aspects of our lives in nature that irritate, sadden, threaten, disturb, or enrage us—our vulnerability to disease and death, to floods, fires, earthquakes, and storms, to random miseries of every sort. Smithson used strong language to express his contempt for this fantasy, and for artists and designers who exploit sentimental interpretations of nature of the sort favored in shelter-magazine landscapes and in the green settings contrived to complement outdoor sculpture, those "graveyards above ground—congealed memories of the past that act as a pretext for reality."[18]

Such ridicule was meant to clear out the house of ideas to make room for a more inclusive interpretation of the world as we understand and experience it. All of Smithson's thinking and writing, every work of art that he made, probed this possibility. Not to replace myth with science, certainly, although he had the same passionate preoccupation with the study of natural systems and structures, particularly entropy and chaos, that drove Thoreau to inventory minute details of the biotic life of Walden Pond; indeed, the house of ideas is also the home of myth. For even though the human animal can know the world in ways that work independently of discursive thought, our communication of every conscious idea depends on symbols— hammered into language, images, sounds, signs—that are the coinage of human exchange. Perhaps knowing that science as much as other narratives is once removed, distanced from the hard quiddity of the material world, contributes to the illusion that we are somehow outside of nature. This is the sense of Smithson's observation that "all thought is subject to Nature. Nature is not subject to our systems."[19] Yet although he spoke constantly in terms of dialectical tensions—site/non-site, chaos/order, reason/derangement—Smithson was not looking for synthesis or resolution, but for contradiction and simultaneity and the deliberate subversion of linear, single-track meanings.

There is an ironic parallel between this aesthetic intention and Robert Venturi's advocacy of a "both/and" rather than "either/or" inclusiveness in architecture and urban design in his 1966 *Complexity and Contradiction in Architecture,* which served as a manifesto for

architectural Postmodernism. But in exposing the utopian and pur-
ist premises of Modernism in order to embrace the messy vitality of
ordinary urban landscapes along with high-art traditions of building, Venturi was still preoccupied with essentially formalist questions of style. For that matter, so is the late-twentieth-century discourse that has emerged from the enthusiasm of many within the
architectural community for critical theory. Although the wordplay
of construction/de(con)struction suggested the sort of dialectic that
might have intrigued Smithson, we can be fairly certain that his earnest materialism—"My work is impure; it is clogged with matter"—
would have abhorred the etherealization, the preciosity of much of
the work and writing associated with that movement.[20]

In searching out mines, dumps, and other places wasted by natural
or industrial processes, Smithson was part of a larger group of environmental artists who either chose sites resistant to conventional
standards of landscape beauty or defied those standards through the
works themselves. One recalls, for example, the early work of Walter
De Maria, George Trakas, and Alice Aycock. Images of De Maria's
much-published *Lightning Field* (1974–77), a grid of four hundred
stainless-steel poles covering more than a square mile of remote New
Mexico desert and designed to attract random strikes, came to represent a body of work dealing with elemental forces—natural or psychic, real or imagined—that had a threatening edge, forcing awareness of realities from which the mundane rituals of daily life usually
shelter us. Trakas frequently used dynamite as a tool in shaping a
landscape, as in his *Union Pass* for the 1977 Documenta VI exhibition in Kassel, Germany. Aycock's memorable *Simple Network of
Underground Wells and Tunnels,* a 1976 installation on a rural site in
New Jersey, allowed those willing to engage the human fear of dark
places to climb down a ladder and crawl through an underground tunnel connecting six concrete wells, illuminated only by light entering
from three wells not capped by a cover of earth.

Art probing these frontiers of the human/nature dialectic was de-
toured not just by the loss of its most compelling theorist but by
the difficulties inherent in adapting it to the programs established by
the Art in Public Places program of the National Endowment for the
Arts (NEA), another initiative begun in the 1960s. Here was exactly
the kind of governmental support of the nation's cultural life that
Olmsted had favored and that many artists, from the New Deal on,
had hoped would democratize patronage and restore art to a central

Walter De Maria, *Lightning Field,* near Quemado, New Mexico, 1974–77. Stainless steel poles (average height: 20 feet, 7½ inches); overall dimensions: 5,280 feet × 3,300 feet. Copyright Dia Art Foundation. Photograph by John Cliett.

place in American education and experience. Even before the recent crises that have endangered all such arts funding, however, it became obvious to artists and also to architects and landscape architects who participated in the collaborative projects favored by the NEA that tough, risky, or otherwise confrontational proposals for landscape and urban projects were problematic if not impossible candidates for funding.

Part of the problem, admittedly, has to do with accommodating within public spaces, especially in cities, work whose aesthetic values derive—no matter how antiquated such categories might now seem—from the sublime or even the picturesque sensibility. In fact, the firestorm of controversy that ended in the 1989 removal, ten years after its commissioning by the General Services Administration, of Richard Serra's *Tilted Arc* from Federal Plaza in lower Manhattan, can be understood in these terms. The 12-foot-high, 120-foot-long, wall-like steel sculpture extending across the plaza was despised as ugly and intrusive by more than a thousand office workers who signed petitions demanding its removal. Even an environmental sculpture as poignant and serenely beautiful as Maya Lin's Vietnam Veterans Memorial occasioned significant public protests early in the 1980s; thinking about death in war in any terms other than those of guts-

and-glory patriotism was deeply disturbing to some Americans. At the least, conflicts such as these demonstrated that the notion of "an American public" was itself an idealized abstraction; there are as many publics, in the end, as there are political, social, and cultural constituencies with agendas to promote. The pressure of that reality inevitably favors inoffensive art, among which works celebrating the beauty and beneficent harmony of the natural world and our happier connections with it are a safe bet.

One of the lasting legacies of the 1960s zeitgeist, however, was the average American's heightened awareness that ecology matters, no matter which side of any given environmental debate makes the most sense. Environmental art that claims to be ecologically correct or expressive of ecological values must inevitably engage those ideas, mythic as well as scientific, that color our preconceptions about nature, in us and "out there."

In some work, an explicit didactic intention—art as a viable and occasionally even entertaining form of ecological education—appears paramount. Helen Mayer Harrison and Newton Harrison's *Survival Piece* series, begun in 1971, explicated various agricultural and aquacultural processes to heighten awareness of our dependence not just on plants, fish, and animals but on specific human technologies toward which we are usually indifferent. Mierle Laderman Ukeles has built her career around a series of works conceived as "maintenance art," vividly dramatic performance pieces and multimedia installations meant to force awareness of the complex physical and social systems that process urban waste. She has been working since 1981 on *Flow City,* a permanent installation at a New York Department of Sanitation facility that moves tons of garbage collected by trucks onto barges for transfer to the aptly named Fresh Kills Landfill in Staten Island. Ukeles is determined to make clear the multiple layers of meaning in her work: "Waste represents the ending of use, it's a metaphor for death, something most of us are afraid to deal with."[21] But she also wants to elicit communal resolve to make radical changes in the social, technological, and natural systems through which waste is processed. We can begin, her work suggests, simply by showing a decent respect for sanitation workers and recognizing the beauty and utility in detritus. Beyond such modest gestures, however, Ukeles wants us to face squarely the need to reduce significantly the volume of waste we produce, and to demand that government at every level help us to recycle as fruitfully and efficiently as nature does.

Art of this sort is deliberately antiromantic and proletarian in comparison with that of artists and landscape architects whose projects, also accommodated under the eco-art rubric, aim to accomplish ecological improvement—restoring wetland vegetation, creating wildlife habitat, cleaning water—without troubling anyone's conscience, much less driving home a moral imperative. In fact, some critics have expressed concern that the land reclamation projects in which artists are now involved might actually reinforce the illusion that we do not need to change how we do business with nature, since a marriage of technology and art may encourage us to fix the appearance of what has been done without addressing the brutal consequences of commercial and industrial operations in which profitability and growth are the bottom-line values. But this is carping, for it implies that doing nothing may be better than doing something, that local and particular actions directed toward the remediation of natural systems are of little value because they will not put off the apocalyptic day of reckoning.

How else will the needed revolution in our thinking and values come about, except through models that capture human imagination and galvanize desires as insistent as those that have produced the institutions and culture, the commerce and industry, of the past five hundred years? The warnings issued by the scientific community, which attempt to raise popular awareness of profound environmental dangers, are clearly not enough. They increase anxiety but cannot of themselves arouse a fierce communal will—the kind of encompassing passion that Henry Adams recognized as having once produced a cohesive medieval culture devoted to the Virgin Mary.[22] Adams compared the power of the Virgin—a moral power that turned belief into action—to that of the mechanical Dynamo of modern times. The art and architecture of the Middle Ages nourished and celebrated reverence for the power and mystery symbolized by the Virgin; shared feelings, more than facts, Adams argued, brought about a profound and creative transformation of Western European culture.

For a radical transformation of contemporary culture to occur, the passions as much as the minds of millions of people must be won over to the cause of saving the earth in ways large and small. Art—music and literature as well as the visual arts, architecture, and landscape architecture—can play a critical role in encouraging this new sensibility, providing that it reaches beyond the educated elites to engage ordinary men and women. We should support the work of artists experimenting with forms and media through which to express the ideas and values of an emerging ecological aesthetic—a propaganda art of sorts, but one that does not require the sacrifice of critical discernment and debate.

A communal undertaking of such enormous complexity begins with personal commitment and demands tolerance for the difficulty of thinking and working communally. Many art projects already seek to achieve social or environmental benefits through the collaboration of artists, scientists, and designers; such projects run counter, of course, to the enduring image of the artist as one who withdraws from society into contemplative solitude. There is great hope in this—in the promise of artists and environmental designers reaching out to a large and popular audience, inspiring us to see with new eyes, to think anew about how we might live with nature, to work with others to achieve the kind of changes Wendell Berry and others see as essential to progress if not survival. Meanwhile, we had better resist the Circe's song of quietists and utopians impatient with the truth that neither we nor the world, in the very nature of things, is, or ought to be, perfect.

2000

Notes

1. Wendell Berry, "In Distrust of Movements," *Orion* 18 (Summer 1999): 17.
2. Ibid., 17–18.
3. Quoted in Louis L. Noble, *The Course of Empire, Voyage of Life and Other Pictures of Thomas Cole* (New York: Cornish, Lamport & Co., 1853), 58–59.
4. Quoted in note 21, Elizabeth Hutton Turner, "Going Home: Geneva, 1933–1938," in *Arthur Dove: A Retrospective,* by Debra Bricker Balkan with William C. Agee and E. H. Turner, catalog (Cambridge, Mass.: Addison Gallery of American Art, Phillips Academy, and MIT Press, 1997), 111.
5. Quoted in Barbara Haskell, *Arthur Dove,* catalog (San Francisco: San Francisco Museum of Art, 1974), 118.
6. Quoted in Laurie Lisle, *Portrait of an Artist: A Biography of Georgia O'Keeffe* (New York: Seaview Books, 1980), 211.
7. Paul Rogat Loeb, *Soul of a Citizen: Living with Conviction in a Cynical Time* (New York: St. Martin's Griffin, 1999).
8. Loeb, "Would the Real Rosa Parks Please Stand Up?" *Atlanta Journal-Constitution,* September 12, 1999, G1.

9. Jedediah Purdy, *For Common Things: Irony, Trust, and Commitment in America Today* (New York: Alfred A. Knopf, 1999).

10. Marshall Sela, "Against Irony," *New York Times Magazine,* September 5, 1999, 58.

11. Quoted in ibid., 58.

12. Quoted in Sam Hunter and John Jacobus, *American Art of the Twentieth Century* (New York: Harry Abrams, 1973), 449.

13. Philip Leider, "How I Spent My Summer Vacation, or Art and Politics in Nevada, Berkeley, San Francisco, and Utah," *ArtForum* 9 (September 1970): 42.

14. Frederick Law Olmsted to Charles L. Brace and Charles L. Elliott, December 1, 1853, Olmsted Papers, Manuscript Division, Library of Congress.

15. Robert Smithson, "Frederick Law Olmsted and the Dialectical Landscape," *ArtForum* 11 (February 1973): 65.

16. John Coplans, "Robert Smithson: The Amarillo Ramp," *ArtForum* 12 (April 1974): 44.

17. Smithson, "Frederick Law Olmsted," 65.

18. Quoted in Lucy R. Lippard, "Breaking Circles: The Politics of Prehistory," in *Robert Smithson: Sculpture,* by Robert Hobbs (Ithaca, N.Y.: Cornell University Press, 1981), 39.

19. Ibid., 32.

20. Ibid.

21. Quoted in Barbara C. Matilsky, *Fragile Ecologies: Contemporary Artists' Interpretations and Solutions,* catalog (New York: Rizzoli, 1992).

22. Henry Adams, *The Education of Henry Adams* (New York: Oxford University Press, 1999 [1918]).

5

Kiss Nature Goodbye:
Marketing the Great Outdoors
John Beardsley

Hats off to commodity culture. The endless quest for new products has spawned another hot-selling hybrid: the not entirely entertaining, but not really educational, simulated "natural" attraction. At the Mall of America, it is called UnderWater World. You pay $10.95, go down the escalator, and enter a dark chamber where synthetic leaves in autumnal tints rustle as you pass. You are in a gloomy boreal forest in the fall, descending a ramp past bubbling brooks and glass-fronted tanks stocked with freshwater fish native to the northern woodlands. At the bottom of the ramp, you step onto a moving walkway and are transported through a 300-foot-long transparent tunnel carved into a 1.2-million-gallon aquarium. All around you are the creatures of a succession of ecosystems: the Minnesota lakes, the Mississippi River, the Gulf of Mexico, and a coral reef. You'll "meet sharks, rays, and other exotic creatures face to face." Sound like fun? Call 1-888-DIVE-TIME (no kidding) for tickets and reservations.

By itself, UnderWater World won't rock the planet. But if not particularly significant, this piece of concocted nature is emblematic of a larger phenomenon. I refer to the growing commodification of nature: the increasingly pervasive commercial trend that views and uses nature as a sales gimmick or marketing strategy, often through the production of replicas or simulations. Commodification through simulation

Rainforest Café, Animal Kingdom Theme Park, Disney World. Photograph by Stuart Franklin/ Magnum.

is most obvious in the "landscapes" of the theme park and the shopping mall. Such places have been widely discussed of late: the malling and theming of the public environment and the prevalence of simulation as a cultural form have elicited skeptical commentary from, among others, philosopher Jean Baudrillard, architectural historian Margaret Crawford, essayist Joan Didion, novelist and semiologist Umberto Eco, design critic Ada Louise Huxtable, and landscape designer Alexander Wilson.[1] Readers of cultural criticism are generally familiar with these writers, and I will not rehearse their arguments. Rather I want to put a different spin on the problem, one attuned especially to the landscape.

Specifically, I am interested in how the commercial context is modifying our conceptions of nature—changing the cultural meanings and values of nature. While many observers focus on the more outrageous examples of malling or theming—the many mutations of Disneyland, Universal City Walk in Los Angles, the various metropolitan simulacra in Las Vegas—I want to explore how nature is packaged for consumption in more ordinary places: first, in our local malls; and second, in nature-based theme parks. The phenomena that are changing—even distorting—traditional conceptions of nature are not limited to a few flagship sites but rather are pervading our surroundings. Almost everywhere we look, whether we see it or not, commodity culture is reconstructing nature.

This is not, however, another requiem for lost wilderness. We humans have always modified our landscapes—sometimes for better, sometimes for worse. Nature shapes culture even as culture inevitably alters nature. Nor is this a lament for some more "authentic" version of nature. The membrane separating simulation and reality is far more porous than we might want to acknowledge. For at least five centuries, since the fifteenth-century Franciscan monk Fra Bernadino Caimi reproduced the shrines of the Holy Land at Sacro Monte in Varallo, Italy, for the benefit of pilgrims unable to travel to Jerusalem, replicas of sacred places, especially caves and holy mountains, have attracted the devout. In the United States, simulations have featured prominently in entertainment landscapes: the 1915 Panama-Pacific International Exposition in San Francisco, to name just one gaudy example, included a scenic railroad whose route featured fabricated elephants, a replica of Yellowstone National Park complete with working geysers, and a mock-up Hopi village constructed by the Santa Fe railroad. But if simulations are not new, they have, in recent decades, become almost ubiquitous, and increasingly they are being used for commercial purposes. And this raises an important question: what does it mean when nature is for sale at the local mall or the downtown boutique? Is this beneficial or harmful? Perhaps it is time to abandon at least some of our ideas about nature—Nature with a capital "N," imagined as independent of culture—and to take the measure of the new "nature" we have been creating.

Why should the commercial landscape matter to designers and those concerned with the designed environment? Partly because there is much room for improvement in this commercial landscape: such places might benefit in interesting ways if the familiar landscape types, the park and the garden, were expanded to encompass the food court, the parking lot, and the flood-control sump hole. It also matters because the commercial landscape is an embodiment of demographic trends that cannot be ignored. Over the past few decades, as suburbia has become more politically and culturally dominant, shopping malls have become one of the centers of our culture. Far more than shopping happens there: malls have usurped many public functions; indeed, they have challenged prevailing ideas of "public space"—malls are ersatz town centers, with police stations, registries of motor vehicles, satellite educational institutions. They have even been venues for weddings (convenient, no doubt, for those last-minute gifts). But, of course, they are not really civic spaces; these privately owned emporiums encourage discreet forms of economic exclusion and social regimentation. As entertainment and tourist landscapes, they are open only to those who can pay. As private places, they limit freedoms of speech and assembly. The Mall of America, for instance, enforces a curfew: those under sixteen must be accompanied by an adult on weekend evenings; everyone under twenty-one must carry government-issued photo identification.

Yet malls are wildly popular and increasingly the locus of our leisure activities. The Mall of America claims to be the "most visited destination" in the United States, annually attracting more than forty million visitors. On the regional scale, Potomac Mills, an outlet mall in northern Virginia that receives twenty-three million visits a year, is now the number-one attraction in that state, outpacing even the Civil War battlefields, Colonial Williamsburg, and amusement parks like Busch Gardens and Paramount's Kings Dominion.[2] Malls feature entertainment and recreation for children and adults, ranging from multiscreen cinemas and video-game arcades at local malls to full-scale enclosed amusement parks like Camp Snoopy at the Mall of America. Almost everywhere, pay-for-play is available in some form: one can play indoor miniature golf, climb synthetic rock walls, face down foes in games of laser tag and in virtual-reality intergalactic battles. Who needs a park—one of those places with grass and trees, playgrounds and benches—when you can while away a Sunday afternoon in a safe, sanitized, and economically segregated simulation of a public space?

How exactly is nature constructed in these new commercial places? When we shop at stores like the Body Shop and the Nature Company, we receive whole sets of messages about nature. At the Body Shop, "natural essences" are linked to physical, emotional, and economic personal fulfillment. The customer here is flattered: the pampering provided by natural oils and aromas is a reward for worldly success; nature here is understood as a source of products that make you feel good rather than as a primary and complex phenomenon for which one bears personal and social responsibility. At the Nature Company, thousands of goods, from stuffed tigers to geodes, from bird feeders and wind chimes to field guides and wildlife videos, have the effect of condensing space and time. The nature presented in this chain store is not localized, dynamic, and differentiated into various ecosystems but homogenized, static, portable, and consumable. The Nature Company implies a kind of global-scale bricolage, a grab bag of concoctions from who knows where, the whole assortment saying little about the actual places from which these products originate. Rather, the goods speak tellingly about those for whom they are marketed— about affluent people who enjoy nature in their spare time, who care about nature and want to be surrounded by things that express their concern. What is really available at these stores, however, is not on the shelves: in a sense, the chief product of the Body Shop and the Nature Company is irony, albeit unintended. These boutiques promise to help us get in touch with nature, but instead they effectively remove us from it—instead of being outside enjoying nature, we are at the mall buying products that express our love of nature.[3]

The larger environment of the mall likewise exploits our affection for nature in order to soothe us in the act of consumption. At most malls, plants are scattered around liberally, especially in the food courts; of course, this indoor profusion compares tellingly with the typical mall exterior, whose gray banality makes the indoor embellishments seem all the more lush. And at some malls—Potomac Mills, for instance—the presence of real plants is supplemented by videotaped nature: large monitors hung at regular intervals in the corridors of Potomac Mills show images of forests and waterfalls. And the entrance to Montgomery Mall, also in suburban Washington, features a large mural of the kind of pastoral scene once characteristic of Maryland. The mural even includes trompe l'oeil binoculars—a device you cannot use to gaze on a landscape that is no longer there.

Nature at the mall can be understood as using the idea of "adjacent attraction," as this has been described by Richard Sennett and Margaret Crawford. Adjacent attraction refers to the phenomenon by which unlike objects, removed from their ordinary circumstances, reinforce each other's value—in other words, if you see something that usually makes you feel good, you might also feel good about whatever happens to be placed next to it. Seeing something unexpected, out of context, makes it seem unfamiliar and therefore exotic, stimulating.[4] Nature at the mall is just such an unexpected phenomenon. The marketing logic is clear: if the customer likes nature, the customer will like the mall and want to shop. Moreover, the demographic group that likes nature most is precisely the affluent group these malls are striving to attract.

In a growing number of malls across the country and around the world, you will now encounter a themed store and restaurant called the Rainforest Café. Billed as "a wild place to shop and eat" and "an environmentally conscious family adventure," the Rainforest Café is one of the more pointed instances of how commercialized mass culture is transforming perceptions of nature. It mixes simulation with reality, entertainment with education, and consumption with conservation. It makes of nature an exuberant and wholly artificial spectacle, even as it aspires to inform customers about actual imperiled ecosystems. The Rainforest Café uses animals for amusement and hopes to protect endangered species. It tempts us to eat and buy and asks us to reduce and recycle. It's imaginative and it's fun, and it's now the best place to experience our confusion about nature and to begin to understand what we ought to do about it.

The first Rainforest Café opened at the Mall of America in 1994; today there are twenty-five (and counting) in the United States and ten overseas. The brainchild of Steven Schussler, an entrepreneur with experience in advertising and restaurants, the concept was developed over some seventeen years. For a while, it existed as a prototype inside Schussler's home in St. Louis Park, Minnesota. Schussler had parrots, toucans, tortoises, an iguana, and a diaper-clad baboon; he claims that his wish to give these pets a cageless environment was what inspired the Rainforest Café. "It became my passion to educate and entertain people about the rain forests, which are the lungs of the world," he said. The café—which combines, says Schussler, "the sophistication of a Warner Bros. store with the animation of Disney and the live animals of a Ringling Brothers/Barnum and Bailey circus"— has proved wildly successful, both popular and remunerative.[5]

In all its diverse locations, the Rainforest Café is much the same, a large shopping and dining space decorated and lit to simulate a fabulous and exotic jungle. The Café at Disney Village is perhaps the most elaborate to date. It is housed in a sixty-five-foot-tall artificial mountain that steams and thunders at regular intervals. Water cascades down the outside, past real and fake plants. Audio-animatronic animals, from butterflies to iguanas to giraffes, beckon to passersby. The cavernous interior is divided into retail and eating areas by a huge, arched, saltwater tank containing simulated coral and real tropical fish. The bar is constructed to look like a gigantic toadstool. Walls are fashioned of faux rock; artificial plants, vines, and Spanish moss are draped from every nook and cranny. Water spills down the walls and mists out from fountains. Fans pump into the air a "rain forest aroma" created by Aveda Corporation from floral extracts. Simulated tropical storms convulse the place every twenty-two minutes, complete with flashes of lightning and booms of thunder. More audio-animatrons perform: elephants trumpet, and gorillas beat their chests. Meanwhile, real macaws and cockatoos are on display for consumers' amusement and edification.

In the end, the Rainforest Café is more jumble than jungle. While purporting to depict the tropical rain forest, it creates a hodgepodge of different ecosystems, from ocean reef to savanna. Stirred into the mix are such non-rainforest animals as zebras, giraffes, and elephants. The mélange of environments makes the Rainforest Café seem less like a simulation than like what Jean Baudrillard describes as a simulacrum: a copy for which no precedent exists.[6] As if that were not confusing enough, the place features a talking banyan tree named Tracy, who delivers the ultimate mind bender: at intervals she mouths exhortations to "reduce, reuse, and recycle," while at other intervals she urges us to try the tasty food or buy the themed merchandise. Conserve, conserve, conserve. Consume, consume, consume. More and more, as the Rainforest Café makes acutely clear, conservation and consumption are two sides of the same cultural currency.

But this simulated cloud has a silver lining. Each Rainforest Café features an animatronic crocodile in a pool into which customers throw coins. The money is collected and donated to groups who work to preserve the rain forest, including the Rainforest Alliance, the Center for Ecosystem Survival, the Rainforest Action Network, and the World Wildlife Fund. Simulated rather than real coral is used in the fish tanks so as not to contribute to the destruction of living reefs. Only line-caught fish are served, and the kitchen tries to avoid buying beef from deforested areas. In 1999, the company became the first national restaurant chain to serve what it calls "bird friendly" shade-grown organic coffee—coffee that does not require the extensive clearing of trees, since it can be grown under the existing canopy. Live animals at the Rainforest Cafés are carefully tended: each café has a full-time curator (some with degrees in ornithology or marine biology) and a trained staff to care for the fish and birds. The birds come not from the wild but from selected domestic breeding programs.

And in 1997, the corporation reportedly spent $175,000 per unit on outreach programs, for instance, taking the parrots to schools and "educating children about the plight of the rain forest."[7]

But none of this conspicuous do-gooding disguises the fact that both the ecological and educational messages at the Rainforest Café are garbled. We can, so the message goes, have it both ways—we can consume and conserve at the same time. This is illogical, to say the least, and possibly deceptive, in that it might comfort some into thinking that consumption and waste are not among our most pressing social and environmental challenges. As at the Nature Company and the Body Shop, the Rainforest Café tries to entertain us, to make us feel good in its approximation of nature, but it does not teach us to be responsible. (In fairness, its promotional material claims only that the chain is "environmentally conscious," not environmentally responsible.) We do not learn about the remote consequences of our consumption—what resources were used, what environments might have been disrupted, what exploited labor might have been employed, where our waste will end up. Under the guise of education and conservation, the Rainforest Café sustains the overconsumption of resources that characterizes middle- and upper-class American life. At the same time, it offers a lesson in the commercial re-creation of nature; as a simulacrum, it exemplifies a new phenomenon in which nature is neither represented nor copied, but replaced by a wholly human concoction.

Malls are only one part of the commercial landscape; the same redefinition of nature is happening at theme parks. Sea World, for example, has recently opened the five-acre Key West World, just 350 miles from the real Key West. It is a miniaturized, sanitized version of the southernmost U.S. city, complete with pastel-shaded knockoffs of local architecture, nonalcoholic margaritas, and performers impersonating colorful characters of Key West, including a sand sculptor ("she's interactive," we are told) and a magician pretending to be W. C. Fields selling swampland. Key West World includes interactive animal exhibits: a stingray lagoon, where kids can feed shrimp to the rays, and an artificial reef with dolphins to feed and pet. The reef features plastic sea fans and fiberglass-reinforced concrete coral; it is gently washed by mechanical waves, which roll back to reveal plastic shells in rocky grottoes. Fake rocks are better than real ones, Sea World's curator told a *Washington Post* reporter, because "real rock doesn't have as much character as the molds."[8]

Sea World is more famous for a show featuring Shamu, the copyrighted killer whale, which combines real-time animal action with a large-screen video documenting the lives of orcas in their natural habitat. The real and the reproduced play simultaneously here, accompanied by a voice-over that allows Sea World's parent company, Anheuser-Busch, to boast that they are being good corporate citizens, researching and protecting endangered species—although they never explain precisely how. They want you just to have fun and relax—multinational capital is minding the environment. From the show itself, however, you would never know that they had learned anything about whales. The exhibit features whale behavior as a form of acrobatics without explaining its function in the wild. One of the most spectacular moments of the show, for example, occurs when the whales launch themselves out of the water and slide at great speed across a platform. What you are not told is that this is a predatory behavior whose purpose is to capture seals. Presumably the message of one lovable sea creature consuming another is deemed too "negative" for the entertainment industry.

I have mixed feelings about nature as presented in places like Sea World. To the extent that the Sea World experience imparts information, I applaud it. But the balance generally tips toward a lowest-common-denominator kind of entertainment, as in a "ride" called Polar Expedition, which involves taking a seat for a simulated flight, shaking around a lot, and pretending to land somewhere in the Arctic. Leaving the flight simulator, you enter a dark, icy landscape at a supposed polar research station, with tanks containing walrus, narwhals, and polar bears. Underlying Polar Expedition in particular and the Sea World experience in general seems to be the assumption that the animals are not sufficiently compelling on their own, that some feeble narrative—or some copyrighted name—has to be fashioned to make them amusing.

Like UnderWater World, Sea World is, finally, neither very entertaining nor very educational. Both represent the burgeoning "infotainment" industry, which attempts to combine the features of an amusement park with the educational mission of the nonprofit zoo, aquarium, or natural history museum. But the bottom line is that these are for-profit entities, and education will always take second place to money. A detailed presentation of wildlife ecology and a reasoned discussion of environmental problems does not sell like the thrill of a pseudo-adventure. Just ask Disney: after painstakingly replicating the botany and zoology of the African savanna at Animal Kingdom, they drag you across the place in pursuit of some fictional poacher.

More complex is the issue of touching and feeding wild animals, as is permitted with dolphins at Key West World. Allowing people to develop psychological bonds with animals perhaps encourages an emotional commitment to the preservation of other species. Meeting animals on a middle ground somewhere between a civilized and a wild place might even be beneficial to the preservation of native habitats. The kinds of simulations encountered in places like Sea World might even be demographically inevitable. As the global population grows and as ecosystems become more fragmented and imperiled, perhaps the best we can do is to leave these places alone. If we can satisfy our urge to experience fragile ecosystems by visiting a simulation, we do less damage to the real thing. On the other hand, if we can manufacture a really convincing and entertaining fake, who will care about the original? Let the real coral reefs die—they will live on as a simulation, courtesy of Anheuser-Busch or Rainforest Café.

One more problem must be raised: whether for consumption, education, or entertainment, commodified nature is only for the affluent. Entrance to Sea World costs $39.95; neither food nor merchandise at the Rainforest Café is inexpensive. The poor have a different relationship to landscape, one governed by scarcity and recycling. Salvage, not consumption, is a conspicuous feature of low-income culture. Infotainment landscapes are another index of the large and growing distance between the haves and have-nots. We may well be witnessing the emergence of three classes of landscape experience: the affluent will make their ecotours to the remaining fragments of pristine habitat; the middle classes will visit simulations; everyone else will inhabit marginal landscapes, salvaging and recycling to survive.

I will conclude with two questions. Should we resist the commodification of nature in the commercial environment? And if so, how? It will be apparent by now that I think we should, for many reasons. The commercial landscape is implicated in some of the most unsettling trends in contemporary culture: the growing gulf between rich and poor, and hardening patterns of social and economic segregation; the tendency to identify people as target groups of consumers rather than as citizens; the transfer of economic resources from the city to the suburb; and the privatization of communal space and the corresponding devaluation of urban public areas as the locus of civic life. The commercial landscape dishes up an ever-changing menu of amusing diversions that hides the real terms of our relationship to the global environment. In the theme park and the local mall, consumption has no consequences. The commercial landscape promotes the misleading notions that we can conserve and consume at once and that our transactions with nature can invariably be safe, agreeable, and problem-free.

Maybe I am hopelessly reactionary, out of touch with cultural imperatives. Maybe the simulated, commodified nature sold to us in the mall and the theme park is the nature we want. Perhaps we would rather live in the realm of reproductions, inhabiting a simulacrum of nature instead of what I still want to call the real world. Perhaps the commitment to biodiversity and to the cultivation of healthy, unpredictable, dynamic, and (dare I say?) beautiful landscapes is merely nostalgic, rendered obsolete as we remake nature entirely in our own image. Simulated nature is certainly a lot less complex and troublesome than the real thing, and our appetite for it now appears boundless. The sheer popularity of simulations demands that we acknowledge, even respect, their cultural importance. Yet I wonder: are we, as consumers, being given what we really want in these commercial landscapes, or are we being sold a bill of phony goods? Commodity producers do not just make products; they manufacture the desire for them. Consumers are never completely passive, but nor are they (we) immune to the seductive powers of marketing. In the absence of now-constant advertising, perhaps consumers would not be as attached to commodified nature as its producers would like to believe.

It is hard to know how to resist. Only the eco-warriors will opt out completely; the rest of us are unlikely to want to make do with much less than we currently consume. Yet I would like to imagine the emergence of a populist environmental politics, a broad-based challenge to the culture of consumption. A certain amount of organized consumer resistance might inspire commodity producers to reformulate their representations of nature, helping us to imagine, as environmental historian William Cronon puts it, "what an ethical, sustainable, *honorable* human place in nature might actually look like."[9] I would also like to imagine enlightened action on the part of corporations that inhabit the spaces of malls and theme parks. A few are beginning to recognize the economic and public relations benefits of increased recycling and diminished resource consumption, but changes in patterns of production and consumption have been few and largely symbolic. We are a long way from leasing such things as appliances, automobiles, and home furnishings and returning them for recycling when we are finished with them, instead of consigning them to the trash heap.

Whether change comes from producers or consumers, we need to make room in the spaces of the theme park and the shopping mall for some alternative narratives and for some dissent from the ideology of consumption. We need to integrate better our natural and social economies; that is, we need to be able to see the connections between the landscapes of production and consumption—to understand the environmental and human costs of waste. We need to hold out for healthy ecosystems in the city and the suburbs; we need to insist that culture—however much it might flirt with simulation—retain a focus on the real world, its genuine problems and possibilities. At the mall or the theme park, what does this mean? Can we imagine a mall that is also a working landscape—that is energy self-sufficient, that treats its own wastewater, and that recycles its own materials? Can we imagine a theme park that is genuinely fun and truly educational and environmentally responsible all at once? I don't see why not. We have created the "nature" we buy and sell in the marketplace; we should certainly be able to change it.

2000

Notes

1. Literature on commercialization and simulation in the public environment includes Jean Baudrillard, *Simulacra and Simulation,* trans. Sheila Faria Glaser (Ann Arbor: University of Michigan Press, 1994); Margaret Crawford, "The World in a Shopping Mall," in *Variations on a Theme Park,* ed. Michael Sorkin (New York: Hill and Wang, 1992); Joan Didion, *The White Album* (New York: Simon and Schuster, 1979); Umberto Eco, *Travels in Hyperreality,* trans. William Weaver (San Diego: Harcourt, 1986); Ada Louise Huxtable, *The Unreal America* (New York: The New Press, 1997); and Alexander Wilson, *The Culture of Nature* (Cambridge: Blackwell, 1992).

2. Visitation figures for the Mall of America are from the 1997–98 "Destination Planning Guide" provided by the public relations office at the Mall of America; for Potomac Mills, from the Virginia Tourism Corporation, conveyed to me in September 1999 by Michelle LaBarbara, director of marketing for Potomac Mills.

3. For a more sustained analysis of shopping for nature, see Jennifer Price, "Looking for Nature at the Mall: A Field Guide to the Nature Company," in *Uncommon Ground: Toward Reinventing Nature,* ed. William Cronon (New York: Norton, 1995), 186–202.

4. Margaret Crawford discusses the principle of adjacent attraction in "The World in a Shopping Mall," 14; Richard Sennett in his *The Fall of Public Man* (New York: Vintage, 1976), 144–45.

5. Quotations from Steven Schussler are from a telephone interview, January 13, 1997.

6. For more on the idea of the simulacrum, see Jean Baudrillard, "The Precession of Simulacra" in *Simulacra and Simulation,* 1–42.

7. Outreach expenditure figures are from Jean Golden, director of public relations for Rainforest Café, Inc.

8. William Booth, "Triumph of the Fake," *Washington Post,* April 14, 1996, A1, 20. For a more sustained analysis of Sea World, see Susan G. Davis, *Spectacular Nature: Corporate Culture and the Sea World Experience* (Berkeley: University of California Press, 1997).

9. William Cronon, "The Trouble with Wilderness," in *Uncommon Ground,* 81.

6

Beyond Wilderness and Lawn
Michael Pollan

My subject is the future of the garden in America. My conviction is that gardening, as a cultural activity, matters deeply, not only to the look of our landscape but also to the wisdom of our thinking about the environment.

When I speak of the future of gardens, I have two things in mind: literal dirt-and-plant gardens, of course, but also the garden as a metaphor or paradigm, as a way of thinking about nature that might help us move beyond the either/or thinking that has historically governed the American approach to the landscape: civilization versus wilderness, culture versus nature, the city versus the country. These oppositions have been particularly fierce and counterproductive in this country and deserve much of the blame for the bankruptcy of our current approach to the environment.

One fact about our culture can frame my argument: the two most important contributions America has made to the world history of landscape are the front lawn and the wilderness preserve. What can one say about such a culture? One conclusion would be that its thinking on the subject of nature is schizophrenic, that this is a culture that cannot decide whether to dominate nature in the name of civilization, or to worship it, untouched, as a means of escape from civilization. More than a century has passed since America invented the front lawn and

Corning, New York. Photograph by David Graham.

the wilderness park, yet these two very different and equally original institutions continue to shape and reflect American thinking about both nature and the garden. I would argue that we cannot address the future of gardening in America—and the future of the larger American landscape—until we have come to terms with (and gotten over) the lawn, on the one side, and the wilderness, on the other.

As the unlikely coexistence of these two contradictory ideas suggests, we tend reflexively to assume that nature and culture are intrinsically opposed, engaged in a kind of zero-sum game in which the gain of one entails the loss of the other. Certainly the American landscape that we have created reflects such dichotomous thinking: some 8 percent of the nation's land has been designated as wilderness, while the remaining 92 percent has been deeded unconditionally to civilization—to the highway, the commercial strip, the suburban development, the parking lot, and, of course, the lawn. The idea of a "middle landscape"—of a place partaking equally of nature and culture, striking a compromise or balance between the two—has received too little attention, with the result that the garden in America has yet to come into its own.

This assertion might seem unfairly dismissive. Certainly there are many beautiful gardens in America, and many gardeners who garden well and seriously. But would anyone argue that American garden design can match, in scope or achievement, American music or painting or literature or even—to name one of our newer arts—American cooking? Of course, even to draw such a comparison will probably strike some as absurd, since our culture does not generally regard gardening as an art form at all. Historically, American gardens have been more utilitarian than aesthetic or sensual. As a result, the United States, which in this century has made large contributions to virtually all of the arts, has produced very few landscape designers who can claim international reputations. Almost the only American landscape artists known internationally are golf course designers, whose talents are in great demand worldwide. Given our infatuation with the lawn, this is not too surprising. But why isn't there a single American garden designer with the international renown of a Robert Trent Jones?

Whether the wilderness ideal or the convention of the front lawn is more to blame for this situation is debatable. But one indisputable fact strikes me as particularly significant: the lawn and the wilderness were "invented" during the same historical moment, in the decade after the Civil War, around 1870. This suggests that these two very different concepts of landscape cannot only coexist but may even be interdependent. In fact, the wilderness lover and the lawn lover probably have more in common with one another than with the American gardener. But before addressing the prospects for the American gardener, I want to address briefly the history of his two adversaries.

Wilderness

On March 1, 1872, President Ulysses S. Grant signed the act that designated more than two million acres in northwestern Wyoming as Yellowstone National Park; thus was created the world's first great wilderness preserve. Grant was responding to a brilliant campaign on behalf of wilderness preservation waged by (among many others) Henry David Thoreau and Frederick Law Olmsted. Why should the peculiar idea of preserving wilderness arise at this time? Clearly, it owed to the fact that the wilderness was disappearing; as early as 1850 visionary Americans began to realize that the frontier was not limitless and that, unless action were taken, no wilderness would be left to protect. America grew rapidly in the period following the Civil War—and so too did the movement to preserve at least a portion of the fast-receding western wilderness. It is remarkable how quickly the movement developed, given that half a century earlier the wilderness had been demonized as worthless, heathen, unregenerate—the haunt of Satan. Of course, the appreciation of wild nature was an invention of the late eighteenth century, of the Romantics—and more specifically, an invention of people who lived in cities. The urbanization of America in the second half of the nineteenth century formed the essential, indispensable context for the creation of the wilderness park—a good example of the mutual interdependence of civilization and wilderness.

From a philosophical perspective, the romance of "undisturbed" land has done much to keep American gardens from attaining the distinction and status of the other arts in this country. Our appreciation of wild land was not, as in the case of the English, primarily aesthetic—it was imbued with moral and spiritual values. The New England transcendentalists regarded the untouched American landscape as sacred. Nature, to Ralph Waldo Emerson and his followers, was the outward symbol of spirit. To alter so spiritual a place, even to garden it, is problematic, verges, in fact, on sacrilege. For how could one presume to improve on what God had made? Emerson himself was an accomplished gardener, but gardening rarely figured in his published writing. This was, I suspect, because he could not reconcile his sense of the sacredness of untouched land with the gardener's faith that the landscape can be improved by cultivation. I am convinced this unresolved conflict forced him into a bit of intellectual dishonesty. It was Emerson who tried to pass off the truly dangerous idea—at least from the gardener's point of view—that there is no such thing as a weed. A weed, he wrote, is simply a plant whose virtues have yet to be discovered. "Weed," in other words, is not a category of nature but a defect of our perception.

Thoreau brought this idea with him to Walden, where it got him into practical and philosophical trouble. As part of his experiment in self-sufficiency, Thoreau planted a cash crop of beans. In general, as observer and naturalist, Thoreau refused to make what he called "invidious distinctions" between different orders of nature—it was all equally wonderful in his eyes, the pond, the mud, even the bugs. But when Thoreau determines to "make the earth say beans instead of grass"—that is, when he begins to garden—he finds that for the first time he has made enemies in nature: the worms, the morning dew, woodchucks, and, of course, weeds. Thoreau describes waging a long and decidedly uncharacteristic "war . . . with weeds, those Trojans who had sun and rain and dew on their side. Daily the beans saw me come to their rescue armed with a hoe, and thin the ranks of their enemies, filling trenches with weedy dead." He now finds himself making "invidious distinctions with his hoe, leveling whole ranks of one species, and sedulously cultivating another."[1]

But weeding and warring with pests wrack Thoreau with guilt, and by the end of the bean field chapter he can't take it any more. He trudges back to the Emersonian fold, renewing his uncritical worship of the wild. "The sun looks on our cultivated fields and on the prairies and forests without distinction," he declares. "Do not these beans grow the woodchucks too? How then can our harvest fail? Shall I not rejoice also at the abundance of the weeds whose seeds are the granaries of the birds?" Unable to square his gardening with his love of nature, Thoreau gives up entirely on the garden—an act with unfortunate consequences not just for the American garden but for American culture in general. Thoreau went on to declare that he would rather live hard by the most dismal swamp than the most beautiful garden. And with that somewhat obnoxious declaration, the garden was effectively banished from American literature.

The irony is that Thoreau was wrong to assume his weeds were more natural or wild than his beans. Apparently he was not aware that many of the weeds he names and praises as native actually came from England with the white man; they were as much the product of human intervention in nature as his beans were. Far from being a symbol of wildness, weeds are, in fact, plants that have evolved to take advantage of people's disturbance of the soil. One of the casualties of our romance of wildness is a certain blindness: we no longer see the landscape accurately, no longer perceive all the changes we have made (not all of which are negative). All too often when we admire a landscape, we assume it is natural—God's work, not man's. Many New Yorkers, perhaps most, have no idea that Central Park is a garden: a designed, man-made landscape. Even people who know about Olmsted and recognize his genius tend to suspend their disbelief and experience the place as "natural," seeking in Central Park the satisfactions of Nature rather than of Art. Historically Americans have tended to experience the great park less as an element of the city, something specific to urban life, than as a temporary, dreamy escape from urban life, an antidote to the city. We can see how even our urban park tradition is founded on the inevitable antagonism of nature and culture, rather than on an attempt to marry the two.

Any culture whose literature takes for granted the moral superiority of wilderness will find it hard to make great gardens. Its energy will be devoted to saving wilderness rather than to making and preserving landscapes. For the same reason, Americans are reluctant to aestheticize nature, to draw distinctions between one plant and another, to proudly leave our mark on the land. And when we do make gardens, we tend to favor gardens that are "wild" or "natural"—but "the wild garden" and "the natural garden" strike me as oxymorons.

A natural or wild garden is one designed to look as though it were not designed. Whether a wildflower meadow, a bog, or a forest, such gardens are typically planted exclusively with native species and designed to banish any mark of human artifice; sometimes they are called "habitat gardens" or, more grandly, the New American Garden (though they are neither New nor American).

There is a strong whiff of moralism behind this movement, not to mention a disturbing streak of antihumanist sentiment. Ken Druse, perhaps the leading popular exponent of the school today, makes clear that the aim of the natural garden is not to please people. "It's no longer good enough to make it pretty," he writes; the goal is to "serve the planet."[2] Obviously, the natural gardeners have not rethought the historic American opposition between culture and nature—between the lawn and what becomes, in their designs, a pseudo-wilderness. They assume we must choose between "making it pretty" and "serving the planet"—between human desire and the needs of nature. I propose that the word *garden* instead be reserved for places that mediate between nature and culture rather than force us to make a choice that is not only impossible but false.

Natural gardeners seem convinced that human artifice in the garden is actually *offensive* to nature. This is an exceedingly peculiar notion. A few years ago, I published an article detailing my first attempt to plant a natural garden. It was a disaster: the weeds quickly triumphed, and the wildflower meadow I envisioned soon degenerated into a close approximation of a vacant lot. I concluded the article with some fairly banal observations about the benefits of planting annuals in rows. Weeding is made easier, of course, but I also found some elemental satisfaction in making a straight line in nature.

I quickly discovered that straight lines in the garden have become controversial in this country. In one of several letters to the editor, a landscape designer from Massachusetts charged that by planting in rows I was behaving "irresponsibly." By promoting even this small degree of horticultural formalism, this critic argued, I was contributing to the degradation of the environment, since gardening "according to existing aesthetic conventions" relied excessively on fertilizers and herbicides. "Nature abhors a straight line," my correspondent claimed, quoting William Kent, the great eighteenth-century English landscape designer.

Natural gardeners have a point insofar as they advocate organic methods. But the formal garden is not inherently less environmentally responsible than a so-called natural garden. A "wild" garden is not intrinsically healthier, or more preferable to nature, than a well-tended parterre. A garden's ecological soundness depends solely on the gardener's methods, not on his aesthetics.

Speculation about what nature does and does not like has inhibited Americans from learning about form in garden, which seems to me a prerequisite to making good ones. Indeed, at its most essential level, gardening is a process of giving form to nature, an activity neither inherently good nor bad; history shows it can be done well, or badly. The forms we use in shaping our land can be subtle, even imperceptible, though I suspect that most of us will fare better with strongly articulated forms—it takes the genius of an Olmsted or a Jens Jensen to make a satisfying garden with less. Most of us who try to create "free-form" gardens make slack, uncompelling spaces. Straight lines are one of the gardeners great tools, and I am convinced that nature couldn't care less whether gardeners plant their annuals in straight lines or meandering drifts.

Lawn

The romance of wilderness is a quintessential American sentiment, but it is one indulged primarily in books or on vacation. The rest of the time, we tend to act in accordance with a very different idea of nature: the idea that the land is ours to dominate, whether in the name of God, during the nation's early days, or later, in the name of Progress. It is astonishing that one culture could give birth to two such antagonistic strains—to both the worship of wilderness and the worship of progress, which usually entails the domination of wilderness. This latter notion, more manifest in the actual landscape than in our writings about it, has been as deleterious to the making of good gardens as has been the wilderness idea. It is far more likely to give us parking lots and shopping centers . . . and lawns.

Anyone who has ever mowed a lawn can appreciate the undeniable pleasure of bringing a heedless landscape under control, however temporarily. But this is not, except perhaps in America, the same thing as gardening. For when we read that gardening is "the number one leisure activity" in this country, we need to remember that the statistics accommodate all those people for whom "gardening" consists exclusively of the pushing, or often driving, of an internal combustion engine over a monoculture of imported grass species.

The ideology of lawns cannot be reduced to the drive to dominate nature, though certainly that is one element of it. The love of closely cropped grass may well be universal, as Thorstein Veblen speculated in *The Theory of the Leisure Class;* it is a reminder of our pastoral roots and perhaps also of our evolutionary origins on the grassy savannas of East Africa.[3] America's *unique* contribution to humankind's ancient love of grass has been, specifically, the large, unfenced patches of lawn in front of our houses—the decidedly odd custom, to quote one authority, "of uniting the front lawns of however many houses there may be on both sides of a street to present an untroubled aspect of expansive green to the passerby."[4] This definition was set forth by the historian Ann Leighton, who concluded after a career spent studying the history of American gardens that the front lawn was our principal contribution to world garden design. How depressing.

The same rapid post–Civil War growth that made wilderness preservation seem imperative also gave us the institution of the front lawn, the birth of which, as near as I can determine, should be dated on or about 1870. At that time several developments—some social and economic, others technological—combined to make the spread of front lawns possible.

First was the movement to the suburbs, then called "borderlands." Before 1870, anyone who lived beyond the city was a farmer, and the yards of farmers were strictly utilitarian. According to the accounts of many visitors from abroad, yards in America prior to 1870 were, to put it bluntly, a mess. The belief that the American landscape has declined and fallen from some pre-industrial pastoral ideal is, in fact, false; the nineteenth-century rural homestead was a ramshackle affair.

Americans rarely gardened, at least ornamentally. The writer William Cobbett, visiting from England, was astonished by the "out-of-door slovenliness" of these homesteads. Each farmer, he wrote, "was content with his shell of boards, while all around him is as barren as the sea beach . . . [even] though there is no English shrub, or flower, which will not grow and flourish here."[5]

All that began to change with the migration to the borderlands. For the first time urban, cosmopolitan people were choosing to live outside of town and to commute, by way of the commuter railroad system then being built. Their homes, also for the first time, were homes in the modern sense: centers of family life from which commerce—and agriculture—have been excluded. These homes were refuges from urban life, which by the late nineteenth century was acquiring a reputation for danger and immorality. The rapidly expanding middle class was coming to believe that a freestanding house surrounded by a patch of land, allowing you to keep one foot in the city and the other in the countryside, was the best way to live.

But how should this new class of suburbanites organize their yards? No useful precedents were at hand. So, as often happens when a new class of affluent consumers in need of guidance arises, a class of confident experts arose as well, proffering timely advice. A generation of talented landscape designers and reformers came forward, from midcentury on, to advise the middle class in its formative landscaping decisions. Most prominent were Frederick Law Olmsted, his partner Calvert Vaux (a transplanted Englishman), Andrew Jackson Downing, and Frank J. Scott, a disciple of Downing's who would prove to be the American lawn's most brilliant propagandist.

These men were seeking the proper model for the new American suburban landscape. Although they were eventually to develop a distinctly American approach, they began, typically, by looking to England, specifically to the English picturesque garden, which, of course, featured gorgeous lawns, the kind that only the English seem able to grow.

Clearly, Americans did not invent lawns per se; they had been popular in England for centuries. But in England lawns were found mainly on estates. The Americans set out to democratize them, cutting the vast, manorial greenswards into quarter-acre slices everyone could afford. (Olmsted's 1868 plan for Riverside, outside Chicago, is a classic example of how the style of an English landscape park could be adapted to an American subdivision.)

The rise of the classic American front lawn awaited three developments, all of which were in place by the early 1870s: the availability of an affordable lawn mower, the invention of barbed wire (to keep animals out of the front yard), and the persuasiveness of an effective propagandist. In 1832 a carpet manufacturer in England named Edwin Budding invented the lawn mower; by 1860 American inventors had perfected it, devising a lightweight mower an individual could manage; and by 1880 this machine was relatively inexpensive. Before the invention in Peoria, Illinois, of barbed wire in 1872, the fencing of livestock was a dubious proposition, and the likelihood was great that one's beautiful front lawn would be trampled by a herd of livestock on the lam. But soon after the mass-marketing of barbed wire, municipal ordinances were being passed penalizing anyone who let livestock wander freely through town.

As for the effective propagandist, the man who did most to advance the cause of the American front lawn—and thus to retard the development of the American garden—was Frank J. Scott, who wrote a best-selling book called *The Art of Beautifying Suburban Home Grounds*. Published in 1870, the book is an ecstatic paean to the beauty and indispensability of the front lawn. "A smooth, closely shaven surface of grass is by far the most essential element of beauty on the grounds of a suburban house," Scott wrote.[6] Unlike the English, who viewed lawns not as ends in themselves but as backdrops for trees and flower beds, and as settings for lawn games, Scott subordinated all other elements of the landscape to the lawn. Shrubs should be planted right up against the house so as not to distract from, or obstruct the view of, the lawn (it was Scott who thereby ignited the very peculiar American passion for foundation planting); flowers were permissible, but they must be restricted to the periphery of the grass. "Let your lawn be your home's velvet robe," he wrote, "and your flowers its not-too-promiscuous decoration."[7] It is clear that his ideas about lawns owe much to puritan attitudes that regarded pure decoration, and ornamental gardening, as morally suspect. Lawns fit well with the old American preference for a plain style.

Scott's most radical departure from old-world practice was to insist upon the individual property owner's responsibility to his neighbors. "It is unchristian," he declared, "to hedge from the sight of others the beauties of nature which it has been our good fortune to create or secure."[8] He railed against fences, which he regarded as selfish and undemocratic—one's lawn should contribute to the collective landscape. Scott elevated an unassuming patch of turf grass into an institution of democracy. The American lawn becomes an egalitarian conceit, implying that there is no need, in Scott's words, "to hedge a lovely garden against the longing eyes of the outside world" because we all occupy the same (middle) class.[9]

The problem here, in my view, is not with the aspirations behind the front lawn. In theory at least, the front lawn is an admirable institution, a noble expression of our sense of community and equality. With our open-faced front lawns, we declare our like-mindedness to our neighbors. And, in fact, lawns *are* one of the minor institutions of our democracy, symbolizing as they do the common landscape that forms the nation. Since there can be no fences breaking up this common landscape, maintenance of the lawn becomes nothing less than a civic obligation. (Indeed, the failure to maintain one's portion of the national lawn—for that is what it is—is in many communities punishable by fine.) Our lawns exist to unite us. It makes sense, too, that in a country whose people are unified by no single race or ethnic background or religion, the land itself—our one great common denominator—should emerge as a crucial vehicle of consensus. And so across a continent of almost unimaginable geographic variety, from the glacial terrain of Maine to the desert of Southern California, we have rolled out a single emerald carpet of lawn.

A noble project, perhaps, but one ultimately at cross purposes with the idea of a garden. Indeed, the custom of the front lawn has done even more than the wilderness ideal to retard the development of gardening in America. For one thing, we have little trouble ignoring the wilderness ideal whenever it suits us; ignoring the convention of the front lawn is much harder, as anyone who has ever neglected mowing for a few weeks well knows. In fact it is doubtful that the promise of the American garden will be realized as long as the lawn continues to rule our yards and minds.

It is important to note that it is not grass per se that is inimical to gardens; indeed, some patch of lawn is essential to many kinds of gardens—the English landscape garden is unimaginable without its great passages of lawn. The problem is specifically the unfenced front lawns, and that problem has both a practical and theoretical dimension.

In practical terms, by ceding our front yards to lawn, we relinquish most of the acreage available to our gardens. Indeed, this space has effectively been condemned by eminent domain, handed over to the community. In fact, when asked, most people will say they regard their front lawns as belonging to the community, while their backyards belong exclusively to themselves.

Because front-lawn convention calls for the elimination of fences, we have rendered all this land unfit for anything but exhibition. Our front yards are simply too public a place to spend time in. Americans rarely venture into their front yards except to maintain them. As one American landscape designer noted, in the 1920s, our lawns are designed for "the admiration of the street."[10] But consider how novel an idea this is: throughout history gardens have usually been thought of as enclosed places—this concept is embedded in the word's etymology. And while great unenclosed gardens (such as Versailles and the English landscape gardens) have been created, these have invariably been so vast in scale that privacy was not an issue. Our lawns might descend from the English picturesque tradition, in which the making of unimpeded views took precedence over the creating of habitable space, but on the scale of suburban development, a "prospect" is not possible without destroying the possibility of usable individual spaces—and of meaningful gardens.

From a philosophical point of view, the very idea of lawns does violence to the fundamental principle of gardening, as expressed by Alexander Pope: "Consult the genius of the place in all." The lawn is imposed on the American landscape with no regard for local geography or climate or history. No true gardener, consulting the genius of the Nevada landscape, or the Florida landscape, or the North Dakota landscape, would ever propose putting a lawn in any of these places, and yet lawns are found in all of these places. If gardening requires give-and-take between the gardener and a piece of land, then putting in a lawn represents instead a process of conquest and obliteration, an imposition—except in a very few places—of an alien idea and even, as it happens, of a set of alien species (for none of the grasses in our lawns are native to this continent).

And last, the culture of the lawn discourages the very habits of mind required to make good gardens. Besides a sensitivity to site and a willingness to compromise with nature, the gardener, to accomplish anything powerful, must be able to approach the land not as a vehicle of social consensus (which by its very nature will discourage innovation) but as an arena for self-expression.

For all these reasons, it will probably take a declaration of independence from the American lawn before we can expect the American garden to flourish.

The front lawn and the wilderness ideal still divide and rule the American landscape and will not be easily overthrown. But American attitudes toward nature are changing, and viewed from one perspective at least, this leaves room for hope. One of the few things we can say with certainty about the next five hundred years of American landscape history is that they will be shaped by a much more acute environmental consciousness—by a pressing awareness that the natural world is in serious trouble and that serious actions are needed to save it. So how will the American garden fare in an age dominated by such an awareness? What about the wilderness ideal? And the front lawn?

It might seem axiomatic that the greater the concern for the environment, the greater the regard for wilderness. But it is becoming clear that attention to wilderness no longer constitutes a sufficient response to the crisis of the environment. True, there are radical environmental groups, like Earth First! which believe that salvation lies in redrawing the borders between nature and culture—in blowing up dams and power lines so that the wilderness might reclaim the land. But an environmentalism dominated by love of wilderness dates to the era of John Muir, and while it has done much to protect a now-sacred 8 percent of American land, it has offered little guidance as to how to manage wisely the remaining 92 percent, where most of us spend most of our time.

We must continue to defend wilderness, but adding *more* land to the wilderness will not solve our most important environmental problems. But even more important, nor will an environmental ethic based on the ideal of wilderness—which is, in fact, the only one we have ever had in this country. About any particular piece of land, the wilderness ethic says: leave it alone. Do nothing. Nature knows best. But this ethic says nothing about all those places we cannot help but alter, all those places that cannot simply be "given back to nature," which today are most places. It is too late in the day to follow Thoreau back into the woods. There are too many of us and not nearly enough woods.

But if salvation does not lie in wilderness, nor is it offered by the aesthetic of the lawn; in fact, the lawn, as both landscape practice and a metaphor for a whole approach to nature, may be insupportable in a time of environmental crisis. Remarkably, the lawn has emerged as an environmental issue in the past few years. More and more Americans are asking whether the price of a perfect lawn—in terms of pesticides, water, and energy—can any longer be justified. The American lawn may well not survive a long period of environmental activism—and no other single development would be more beneficial for the American garden. For as soon as an American decides to rip out a lawn, he or she becomes, perforce, a gardener, someone who must ask the gardener's questions: What is right for this place? What do *I* want here? How might I go about creating a pleasing outdoor space on this site? How can I make use of nature here without abusing it?

The answers to these questions will be as different as the people posing them and the places where they are posed. For as soon as people start to think like gardeners, they begin to devise individual and local answers. In all likelihood, post-lawn America would not have a single national style; we are too heterogeneous a people, and our geography and climate too various, to support a single national style. And that, after all, has been the lesson of the American lawn: imposing the same front yard in Tampa and Bethesda and Reno and Albany exacts too steep an environmental price. Undoubtedly lawns will survive in some places (such as the cool, damp Northwest, where the "genius" of the place may well accommodate them), but the American front yard will someday be entirely different things in Sausalito and White Plains and Fort Worth. Those who are intent on establishing a "New American Garden" may judge this a loss and balk at giving up the idea of a single national landscape style. But as valuable as unifying national institutions may be, nature is a poor place to try to establish them. However one may judge multiculturalism, multi-*horti*culturalism is an environmental necessity. The New American Garden must be plural.

But even if an age of environmentalism does not attack the lawn head on, it would still bode well for the garden in America. The decline of the lawn may be gradual and piecemeal and even inadvertent, as gardens gradually expand into the territory of the lawn, one square foot at a time. To put this another way: to think environmentally is to find reasons to garden. Growing one's food is the best way to assure its purity. Composting, which should be numbered among the acts of gardening, is an excellent way to lighten a household's burden on the local landfill. And gardens can reduce our dependence on distant sources not only of food but also of energy, technology, and even entertainment. If Americans still require a moral and utilitarian rationale to put hoe to ground, the next several years are certain to supply plenty of unassailable, even righteous ones.

So I am optimistic about the American garden—or at least about the proliferation of gardens in America. As the environment take its necessary and inevitable place in our attention, the reasons to garden will become increasingly compelling, and the reason to maintain lawns correspondingly less so. Whether there will be a flowering of great gardens in America is another question, but here too there is a reason to be optimistic, one that may seem at first entirely off the point: almost overnight, Americans have invented a distinctive and accomplished cuisine.

Only a few years ago American cooking was no better than provincial Britain's or Germany's: unimaginative, heavy, relentlessly utilitarian. (Talk about the plain style!) Our recent culinary revolution suggests that a corner of the culture formerly neglected or disdained may suddenly become the focus and beneficiary of the kind of sustained attention and cultural support that makes genuine, original achievement possible. To take this analogy even further, my guess is that the same radical cosmopolitanism that today distinguishes American cuisine—its willingness to draw from a dozen different national traditions, combining them in never-before-seen combinations—will someday define the New American Garden.

If this analogy seems far-fetched—the lawn giving way to the mixed border the way the meatloaf has given way to the shiitake mushroom and goat cheese pizza—consider for a moment that the preconditions for a brilliant cuisine and a brilliant garden are so similar: both require the artful intermingling of nature and culture. "Cookery," the poet Frederick Turner has written, "transforms raw nature into the substance of human communion, routinely and without fuss transubstantiating matter into mind."[11] Couldn't much the same be said about the making of gardens? Perhaps the old puritan antagonism between nature and culture is at last relaxing its hold on us. If we are finally willing to sanction the mingling of these long-warring terms on our dinner plates, then why not also in our yards?

That would be very good news for the quality of our gardens and also, in turn, for the quality of our thinking about the environment. For if environmentalism is likely to be a boon to the American garden, gardening could be a boon to environmentalism, a movement which, as I have suggested, stands in need of some new ways of thinking about nature. The garden is as good a place to look as any. Gardens by themselves obviously cannot right our relationship to nature, but the habits of thought they foster can take us a long way in that direction—can even suggest the lineaments of a new environmental ethic that might help us in situations where the wilderness ethic is silent or unhelpful. Gardening tutors us in nature's ways, fostering an ethic of respect for the land. Gardens instruct us in the particularities of place. Gardens also teach the necessary, if still rather un-American, lesson that nature and culture can be reconciled, that it is possible to find some middle ground between the wilderness and the lawn—a third way into the landscape. This, finally, is the best reason we have to be optimistic about the garden's prospects in America: we need the garden, and the garden's ethic, too much today for it not to flourish.

1998

Notes

1. Henry David Thoreau, *Walden* (New York: Macmillan, 1962 [1854]), 117–25.

2. Ken Druse, *The Natural Habitat Garden* (New York: Clarkson Potter, 1994), 9.

3. Thorstein Veblen, *The Theory of the Leisure Class* (New York: Viking Press, 1967 [1899]), 134.

4. Ann Leighton, *American Gardens of the Nineteenth Century* (Amherst: University of Massachusetts Press, 1987), 249.

5. William Cobbett, quoted in John Stilgoe, *Borderland: Origins of the American Suburb, 1820–1939* (New Haven, Conn.: Yale University Press, 1988), 71.

6. Frank J. Scott, excerpted in *The American Gardener: A Sampler,* ed. Allen Lacy (New York: Farrar Straus & Giroux, 1988), 317.

7. Ibid., 321.

8. Ibid., 322.

9. Ibid., 323.

10. Grace Tabor, *Come into the Garden,* excerpted in Lacy, *The American Gardener,* 339.

11. Frederick Turner, "Cultivating the American Garden: Toward a Secular View of Nature," in *Harper's Magazine,* August 1985, 52.

II | Desiging (for) Nature

7

Nature Used and Abused: Politics and Rhetoric in American Preservation and Conservation

Rossana Vaccarino

The preservation and conservation of nature may be seen as both a goal of action and the subject of discourse, that is, as a "text" constantly produced, deconstructed, and reinterpreted in human values and conceptions. Like nature itself, conservation and preservation are inseparable from the specialized language that we use to talk about them; they are associated with narratives that are rarely neutral but are instead framed by ideologies with social and epistemological dimensions. Most important, both terms are used strategically by politically diverse factions—progressive and conservative alike—who wish to mediate and mask power relations or to justify forms of land use and control. Throughout the twentieth century, concepts of nature preservation, reclamation, and ecology have been appropriated by public and private institutions in their attempts to legitimize, via environmental rhetoric, discriminatory economic and political practices. I would like to explore here the rhetorical strategies adopted in nature conservation and preservation, as well as the origins of environmental conflict in the United States and their ramifications in the current debate.

Utilitarianism and the Rhetoric of "Wise Use"

The ideological roots of contemporary American environmentalism are usually traced to the conservationist-preservationist debate at the turn of the twentieth century, an era that saw the closing of the frontier and a dawning awareness of the alarming exploitation of the nation's land, water, and forests. During this period, forester Gifford Pinchot began to advocate the "wise use" of natural resources; his rhetoric, which would become influential, stood in contrast to naturalist John Muir's transcendentalist discourse of nature preservation. Together these perspectives recapitulated the ambivalence of the European colonizers in the New World, of the settlers who dreamed of obtaining both material plenty *and* spiritual fulfillment from nature. Muir called himself a "conservationist" but eventually opted for the appellation "preservationist"—this in reaction to Pinchot's appropriation of "conservation" to mean not "resource protection" but efficient "resource development."[1]

Early preservationists were mainly affluent white men, either landed naturalists or associations of hikers, campers, mountain climbers, hunters, and fishermen who saw the wilderness as a place for aesthetic pleasure, sport, hobbies, and recreation. As reporter and editor Mark Dowie has written, "Their organizations were clubby and exclusive: some admitted new members only if sponsored by existing members. The Sierra Club, for years a private mountaineering club, required two sponsors."[2] By the mid-twentieth century the organizations had become less exclusive, but their aims remained limited and conservative.

Pinchot, who had studied forest management in Europe, was a member of the Boone and Crockett Club, an elite hunting and wildlife protection association. There he met Theodore Roosevelt, who, under Pinchot's influence, would later proclaim that "forest protection is a means to increase and sustain the resources of our country and the industries which depend upon them. The preservation of our forests is an imperative business necessity."[3] The historical context of this declaration is revealing: when, in the late nineteenth century, Congress was setting aside federal forest "reserves," most of the first-growth eastern forests had been cleared, and roughly half of the publicly owned forest lands of the West had been transferred to private ownership, mainly to railroad, timber, and mining companies and to homesteaders. The forests thus "protected" from logging were nearly always those the timber companies had rejected due to lack of easy access or good-quality trees. In 1905, Congress created the U.S. Forest Service, whose mandate was to manage the forestlands; Pinchot was appointed its first head.

The compelling rhetoric of Pinchot and his followers was focused on the idea that American forests could be saved from rapid depletion only if managed scientifically—"used wisely." The forest was seen as a form of capital to be invested for future return: trees could be cut for economic gain, and the forest preserved at the same time.[4] This thinking was informed by both the utilitarian ideology of the Progressive Era and the growing science of agronomy. Not coincidentally, the Forest Service was strategically placed within the Department of Agriculture. Moreover, since forest trees were seen

Theodore Roosevelt *(left)* with John Muir, founder of the Sierra Club, at Yosemite Point, California, 1903. Copyright Bettman/CORBIS.

as a crop, they could theoretically be harvested, replanted, and harvested again indefinitely under "sustained yield" management.[5] This metaphor of the "accumulation of forest capital" appealed to the conservative interests of large forest landowners.

The practice of sustained yield forestry became the model for forest management in North America, and it remained so until the late 1980s, when the Forest Service shifted to "ecosystem management," which aims to sustain the health of the forest rather than the wealth of industrialists and landowners. This change was undoubtedly influenced by the timber policies of the George H. W. Bush and Clinton administrations and by pressure from various environmental organizations. What was crucial in this change, though, was the advice of new "experts" from the now established science of ecology.

Pinchot believed that the forest should be managed and used efficiently for the good of *all* society, and he was convinced that this was possible only through the executive control of a national program of progressive reform based on the advice of scientifically trained specialists.[6] For Pinchot, technicians such as foresters, hydraulic engineers, or agronomists, rather than legislators, were the natural leaders of forest management. Unfortunately, in the end, Pinchot's social vision was easily diverted by those who wished to promote economic gain; it was, in fact, appropriated by the very people and entities considered responsible for endangering the forests.

Indeed, by the 1920s and 1930s, the language of conservation was embraced by resource-based industries and other private interest groups, all of whom found the rhetorical notion of the "efficient development of physical resources" well suited to marketing and business strategies. In turn, the Forest Service and the Bureau of Reclamation, under the influence of industry associations, began to drift away from their original mission and increasingly to advocate for particular private interests, including landowners and the timber industry. But it was not until after the Second World War that a "mutually beneficial" triangle was established, the corners of which were Forest Service officials, congressional lawmakers, and timber barons—the hegemonic triumvirate of science, government, and business. And, not coincidentally, it was during this period that logging was made legal in national forests, and those trees that had so eagerly been "reserved" by conservation efforts were made available for the exploding postwar lumber market.[7]

By the 1960s, clear-cutting had been naturalized by government

Clear-cut slopes, Klamath River Basin, California, 1992, from the Farewell, Promised Land Project. Photograph by Robert Dawson.

officials as a "restorative" practice that would allow "thrifty" young forests to replace great old ones. The ensuing debates are well known. Scientific experts found in the "overstocked" forests myriad species, such as the spotted owl, whose survival depended on what some viewed as expendable, "over-mature" trees. Others began to question the "sustained yield" and "wise use" practices by showing that the young "thrifty" forests were not regrowing productively and that harvest rates had been way too high. A counteroffensive arose in the late 1980s, attempting to justify unsustainable logging practices with a cloud of politically correct ecological jargon. Suddenly all national forests were found to be "sick." Trees were "overcrowded" or stressed by fire or disease. A logging "treatment" was urgently needed to "save" the public forests and "restore their health." "Salvage logging" or the need to "log" the forest in order to "save" the remainder of it was a convenient strategy against the environmentalist lawsuits that had by then made traditional logging practices nearly impossible on public lands. This logic was used to conceal the appetites of the forest officials, lawmakers, and the timber industry. One might well wonder what was left of the "wise use" and the "future returns" ideas that Pinchot had so passionately advocated.

Individualism and the Rhetoric of "Setting Apart"

If "utility" was the motto of the early conservationists, the "setting apart of nature" from urban and industrial influences became the mission of the wilderness movement led by John Muir and the Sierra Club. This preservationist movement reflected various attitudes and approaches including nationalism (nature seen as a national treasure analogous to the cathedrals and museums of the Old World), commercialism (wilderness perceived as a commodity for tourism and recreation), spiritualism (wilderness experienced as an Arcadian counterpoint to industrialization, a romantic or sacred refuge from the evils of civilization), biocentric ethics (nature, or the realm of the nonhuman, valued for its own sake), and an elite aestheticism (nature viewed as scenic experience or image, especially for those presumed capable of appreciating its beauty and sublimity).[8] Wilderness advocacy in the United States dates back to the early nineteenth century, but only during the century's closing years did it become a major issue.[9] According to the preservationists, protection of the wilderness—or its remnants—was equivalent to the protection of the nation's most sacred myth of origin: free land in wild nature. To them, wilderness was "the last bastion of rugged individualism."[10]

Muir was a mountaineer-naturalist, a wilderness trekker, and a prolific and popular essayist who valued wild nature primarily as a spiritual and aesthetic experience. His literary and philosophical ideas were deeply influenced by Ralph Waldo Emerson and Henry David Thoreau, by poet William Cullen Bryant and naturalist George Perkins Marsh. From these influences and from that of his evangelist father, Muir invented a new version of "frontier religion," one based on an ascetic "separation from human corruption" through "redemption in the wild," in a wilderness that was, for him, at one with the divine.[11]

Muir's rhetoric reinforced the romantic idea of nature as the source of inspiration for the solitary visionary. He saw wilderness as a place of private experience; he never imagined it as possessing any public or collective dimension, or as an appropriate place for human habitation. Social engagement, he believed, was possible only within urban-industrial civilization. To be sure, Muir disliked urban life; he distinguished between the urban "lowland" and the wilderness "high ground." The preservationists who embraced this credo were often upper-class sportsmen and privileged amateurs who sought to escape the hectic, "unnatural" conditions of city life. Not surprisingly, the notion of the ascetic detachment of the "exceptional individual" was suited to an elite that wanted to separate itself from "ordinary citizens" and avoid the social and environmental problems associated with the industrial city. In 1872, when Yellowstone was made a national "preserve" and public monument by an act of Congress, this idea of the wilderness as a "distinctive" and wondrous place became officially established. Rhetorically at least, from then on, the wilderness preserve, with all its contradictions, became the organizing myth of the American national parks.[12]

After the Second World War, rising living standards enabled more people to afford leisure experiences, including vacations in national parks. A new economic argument that emphasized "making a business of scenery" began to inform the policy of the parks; a new era in the use of wilderness thus began. Greater access to the parks was needed to stimulate and accommodate the expanding tourist trade (which was also spurred by the construction of the Interstate Highway System); movement through the parks—by people and cars—had to be managed, and any disturbances to the parks had to be rectified. Indeed, the parks' managers devoted much effort to attempting to erase the traces of their human visitors. The big task was to keep people from herding together—to allow visitors to experience the increasingly crowded parks as "wild" and "solitary" places. The irony, of course, was that this wilderness came more and more to include the very civilization its visitors sought to escape.

Today, more than ever, wilderness is a commodity whose "sale" is helped by resort and ecotourist destinations, travel packages, recreational vehicles, and retailers like the Nature Company. In fact, the past twenty-five years have witnessed the rise of an industry of hundreds of organizations that sell wilderness experience vacations. The identification with nature thus promoted by these programs is distinctly utilitarian in tone: remote protected areas of the country are advertised as places for healing the human spirit and "developing" or "discovering" confidence and self-esteem. The rhetoric of the preservationist ethic—of nature lovers finding spiritual fulfillment and leaving no trace behind—is employed to justify the use of wilderness for apparently laudable human purposes. A program such as Wilderness Discovery, for instance, offers socially or educationally disadvantaged youth—unemployed, poor, homeless, or victimized by drug or alcohol abuse—the chance, through the rigors and inspirations of camping out, to develop inner strength.[13] Clearly such intended uplifting is little more than the equivalent of a Band-Aid on broad and deep wounds. Indeed, the rhetoric of the "discovery of the inner self" through the discovery of wild nature and the glorification of a hermit lifestyle can even be seen as reinforcing the status quo: action can remain private and motivated by commerce rather than government or public policy.

The Hidden Nature of Conservation and Preservation

The intrinsic aim of both the "resourcist" and "protectionist" rhetoric is to win public approbation through the use of authoritative language; such language is used to build constituencies and achieve consent and identification, rather than to provoke confrontation, with particular environmental policies.[14] During moments of calamity or impasse—such as the period that saw the closing of the frontier—people are more easily allured by appeals based on fear. In the recent history of the Forest Service, for instance, huge amounts of "salvaged timber" from arson fires or what have been diagnosed as "bug-infested" federal forests were sold to the timber industry at bargain prices or at a loss. The government labeled this action an "emergency measure" and presented it as an ecologically correct strategy to "prevent forest fires"—a "treatment" to remove "fuel loads." Despite the rhetoric, however, the operation was hardly intended to be ecological, since it granted timber companies permission to bring heavy equipment to highly erosive and ecologically sensitive hillsides where logging would otherwise be prohibited. Nor did this action prevent fires; arson has since increased exponentially, given the incentive to "salvage" more timber. Moreover, salvage bills have became a normal rather than emergency measure for logging in federal preserves, and several such bills have authorized the sale of "associated" healthy trees along with the dead or dying ones—a clever way to make any tree in the proximity of the "emergency area" qualify as "salvage" and thus sell it at a bargain. Taxpayers, who theoretically own these trees, have lost much money in the process.[15]

As modern American history has shown, consent or collusion with conservationist or preservationist issues has usually been achieved by avoiding any direct challenge to established institutions and ideologies, and any overt questioning of the Western values of progress, economic growth, and consumerism. This soft position goes hand in hand with traditional liberalism and its nonconfrontational tactics and with the conservative and reformist nature of American government. What is remarkable, however, is that the kind of environmental rhetoric discussed here, purified of extreme or subversive facts and ideas, has become resilient enough to be adopted, usually after modification, by those with radically different political tendencies and ideologies. Hence we find a similar environmental eloquence among disparate groups from the political Left and Right—animal rights organizations, social justice activists, humanists, Disney, media corporations, and others.

The naturalization or induced public acceptance of a conflict or event as "inevitable" and its justification in ecological, aesthetic, or technical terms that exclude the political are other strategies to achieve consensus. In this respect, environmentalists are still described by many as minorities with special interests, an organized "resistance" that enters the debate from the "margins of society" with little understanding of the practical needs of a working community. As seen in the disputes between the "tree people" and the lumber industry in the American West, the conflict between environmentalists (preservationists) and those who favor development (conservationists and "resource developmentalists") is usually depicted as irresolvable. The "inevitable" struggle, often compared with the metaphysical polarity between humans and nonhuman nature, is accepted without question. The modern dualism that sees nature as an external, nonhuman reality—as the *object* of scientific study, aesthetic inquiry, or environmental protection—is thus perpetuated, justifying the need and the right to exploit the nonhuman world for human use. In short, it is a move to reduce nature to a material or spiritual resource able to sustain present as well as future human life.

A further illustration of this naturalization is the rhetoric in which the American national parks are "saved" as wilderness for the "American people." Set apart in apparent innocence, these parks become symbols of what Euro-American preservationists have fought for as their origins. Yet, of course, these lands have been inhabited since the last ice age. The American wilderness—set aside right after the final wars with Native Americans, who were exterminated or evicted onto reservations—was not "virgin," "free," or "empty" of human beings.[16] Nature had to be emptied out—in people's consciousness first and then in practical terms—to allow for both the occupation of (the white) man and the "preservation" of nature.

The half-hidden agenda of rhetorical persuasion is control. This is marshaled through "specialists" who attempt to mediate and mask power and to subvert or neutralize the sociopolitical dimensions always in play. This covert demagogy was already endorsed by the Theodore Roosevelt administration around the turn of the past century, when the conservation movement was still young. Under the influence of Pinchot, the government sought the advice of scientifically trained experts to enable "rational planning" for the development and use of natural resources. In fact, the technocratic control and manipulation of information under the rubric of the "objectivity of science" were a means of centralizing authority.[17] The rhetoric of "national efficiency" became a cover for an unspoken mandate to neutralize local needs. In effect, the public, lacking the information and influence to question the "scientific evidence" of experts, was not incorporated into the decision-making process. Even today, the strategy of silencing local needs and discouraging public participation through the testimony of "experts" underlies not only many operations of the corporate media and global market economies but also the technocratic character of mainstream environmental organizations and, directly or indirectly, the federal government. Along these lines, it is worth recalling the controversial and much discussed mechanism of the "environmental impact statement," which serves the interests and reflects the power relations of science, government, and industry, and which incorporates the perspective of the general public only in token ways.

Special interest control can also be achieved through the enforcement of laws and policies. Zoning regulations, for instance, increasingly adopt key aspects of conservationist and preservationist rhetoric. In countless examples across the country, zoning "for nature" has become a device to "zone apart" citizens of different income levels. Within a general cultural phenomenon of suburban and exurban escape, with its utopian echoes of an agrarian, pastoral past, the "nature preserve" and conservation area often become the most "exclusive" landscapes for the rich to inhabit at the exclusion of "undesirable" people or densities—in this case under the pretext of defending nature and allowing only token visitors in the restricted zones. The irony is that while the wealthy are sometimes the major actors and promoters of preservation at home, they are often the owners of companies and corporations that exploit and despoil the land and atmosphere elsewhere.

Throughout, the paradox in the politics of nature protection is that preservationist and conservationist discourses are easily co-opted to promote development and other antienvironmental agendas. The oxymorons of "wise use" and "efficient development" of natural resources have been the vehicles of a rhetoric that suited both pro-tectionist and prodevelopment agendas. These slogans, in the end, were appropriated to promote the commercial ideology and/or eco-nomic benefit of particular interest groups rather than to protect the broader rights and needs of society. The same strategy occurs today in the rhetoric of "sustainable development," which has been used by environmental organizations and global economies alike. The con-flict between development and protection is thus neutralized—the euphemism reassures us that we can eat our cake and have it too. While development is made "sustainable"—able to be continued— the opening of new frontiers in the Third World is facilitated through the "participation" of other cultures in the global economy, capitalist models of progress, and resource exploitation. The likely result is multinational corporate "imperialism" on an unprecedented scale.[18]

Reframing Nature Protection

In contemporary environmental discourse, the preservation and con-servation of nature occur in a political arena, with elite sectors con-tinually seeking local or global control, freeing themselves from ac-countability or the responsibility to address democratic imperatives of public participation and social justice. The environmental move-ment that grew in the early 1960s from the old-line conservationist groups is dominated by a few large national organizations, includ-ing the Sierra Club, the Wilderness Society, the National Audubon Society, the National Wildlife Federation, and the National Parks and Conservation Association. Most of these organizations are centered in Washington, D.C., are reformist in their political strategies, and are heavily staffed by lawyers and MBAs trained to enter political debates with the federal government and corporations. By gaining access to the sources of power and by embracing the rhetoric and the nonconfrontational strategies of their opposition, these organizations risk becoming class-bound interest groups. In this mainstream debate, environmentalism has remained conservative and narrow, equated in the public mind with the protection of scenic wilderness, old-growth forests, or cuddly wildlife.

Environmental politics are now set, however, in an increasingly diversified and informed society, one that encourages shared responsibilities and consensual approaches for long-term solutions to the deepening global ecological crisis. Appeals for endangered species or habitat protection are being put in a broader perspective that takes into account the contributions of a wider range of stakeholders and an array of global policy problems. Significant changes in both national and international institutions and policies have taken place since the Conference on the Human Environment in Stockholm in 1972 and even since the Conference on Environment and Development (or Earth Summit) in Rio de Janeiro in 1992. The new challenge is to establish both local and global environmental governance, connecting ecology to economic and social issues such as poverty, consumerism, public health, demographics, and housing. The range of environmental concerns has broadened beyond deforestation or the loss of "wilderness" to include climate change, land-use planning, desertification, loss of biodiversity, marine pollution, and the disposal of toxic chemicals and radioactive materials. Agenda 21, the Biodiversity Convention, and the Framework Convention on Climate Change, adopted at the Earth Summit in Rio, remain examples of broad-based agreement on both a vision and a blueprint for action, involving the direct contribution of women's groups, indigenous people, business and industry leaders, parliamentarians, labor forces, and thousands of nongovernmental organizations. Together these initiatives form a manifesto for a shift toward participatory, pluralistic environmentalism.

Within this increasing democratization, much remains to be learned and tested, but there are reasons for optimism. The hundreds of citizens who have formed grassroots organizations are no longer expressing parochial, utopian, or single-issue radicalism. Bottom-up organized protest has been central to the environmental effort since the mid-1960s and is growing as a legitimate alternative to the polite activism and legislative/litigation strategies of the preservationist and conservationist movements.[19] This radical movement has been able to react to and confront closed-door decision-making processes, even in developing countries where financial resources, education, and access to government are rare or absent. More important, a common agenda unites this diverse grassroots environmentalism: a call for equity in the distribution and use of resources, freedom from the dangerous effects of environmental degradation, and environmental justice in the fulfillment of basic needs for human health and autonomy.

Increased access to information through television, public lectures, radio, alternative presses, and especially the Internet will significantly expand the number of players willing to engage the political and social dimensions of environmentalism directly. The empowerment provided by direct access to information might serve to inspire and mobilize a wide range of social sectors and to leverage their role in environmental governance. It seems likely that people will emerge from their private lives and search for those skills and tools that enable them to take control, organize, and evolve. At the same time, a more effective and innovative process for conflict resolution than legal battling may develop. A nature conservation debate about "Habitat Conservation Plans" is even now enlivening many Internet sites.[20] Stay tuned.

2000

Notes

1. The designation "environmentalist" and the notion of protecting both human and natural environments came into common usage only after the 1964 publication of Rachel Carson's *Silent Spring,* when most conservationists were still preoccupied with preserving wilderness and protecting wildlife.

2. Mark Dowie, *Losing Ground: American Environmentalism at the Close of the Twentieth Century* (Cambridge, Mass.: MIT Press, 1995), 2–3.

3. The Roosevelt speech is cited in Gifford Pinchot, *Breaking New Ground* (New York: Harcourt, Brace, 1947), 190.

4. Gifford Pinchot, *A Primer of Forestry: Practical Forestry* (Washington, D.C.: U.S. Department of Agriculture, Bureau of Forestry, Bulletin 24), 41.

5. The instrumentalist view of nature as a source of value is obviously not new; it dates back to Enlightenment philosophy and capitalist modes of production in the eighteenth century.

6. Pinchot's social vision was not followed in the Forest Service, but during the New Deal, under the leadership of Robert Marshall, a short-lived critique of development pressures on forestlands was put forward and a link

between social justice and wilderness protection was established. See Robert Marshall, *The People's Forests* (New York: H. Smith and R. Haas, 1993).

7. For more details on the events, see Paul Roberts, "The Federal Chain-Saw Massacre," *Harper's Magazine,* June 1997, 37–51.

8. I am expanding here on the classification provided by Robert Gottlieb in *Forcing the Spring: The Transformation of the American Environmental Movement* (Washington, D.C.: Island Press, 1993), 27.

9. The original roots of wilderness as a human construct trace back to the emergence of agriculture as a social pattern in human life. See Roderick Nash, *Wilderness and the American Mind* (New Haven, Conn.: Yale University Press, 1982).

10. William Cronon, "The Trouble with Wilderness: or, Getting Back to the Wrong Nature," in *Uncommon Ground,* ed. William Cronon (New York: W. W. Norton & Co., 1996), 77.

11. On the nature theology of John Muir, see Donald Worster, *The Wealth of Nature: Environmental History and the Ecological Imagination* (New York: Oxford University Press, 1993), 184–220; and Max Oelschlaeger, *The Idea of Wilderness* (New Haven, Conn.: Yale University Press, 1991), 172–204.

12. At the beginning, the national parks relied for access on railroad development, which was geared to upper-class tourism and became the primary source of capital for expensive new facilities, grand hotels, and other concessions. Early national parks were often posted "whites only."

13. See Marla Kale, "Wilderness and the Human Spirit," *American Forests,* January/February 1995, 39.

14. For the concept of identification, see Kenneth Burke, *A Rhetoric of Motives* (Berkeley: University of California Press, 1969), cited in *Ecospeak,* by M. Jimmie Killingsworth and Jacqueline. S. Palmer (Carbondale: Southern Illinois University Press, 1992), 7, 23.

15. Roberts, "The Federal Chain-Saw Massacre," 48–49.

16. There are many sources on this topic. See Cronon, "The Trouble with Wilderness," 79; and Dowie, *Losing Ground,* 11.

17. See Samuel Hays, *Conservation and the Gospel of Efficiency* (Cambridge, Mass.: Harvard University Press, 1959).

18. One of the most eloquent advocates of this view is Vandana Shiva. See Vandana Shiva, *Monoculture of the Mind* (London: Zed Books, 1993); and *Biopiracy: The Plunder of Nature and Knowledge* (Boston: South End Press, 1997).

19. Within the movement for forest preservation, an example is the group of citizens who have formed a "movement-within-a-movement" to stop the clear-cutting of ancient trees. Fighting the logging industry one timber sale at a time, they have saved more trees in the past decade than have been saved by federal legislation. See Dowie, *Losing Ground,* 261.

20. Habitat Conservation Plans (HCPs) were proposed in 1982 as an amendment to the Endangered Species Act of 1973. The amendment was perceived as a "win-win" solution, a way to preserve endangered species habitat and to satisfy development interests. Since the start of the Clinton administration, over 320 HCPs have been approved or are being considered for approval, with increasing trends toward regional, multispecies plans involving thousands or hundreds of thousands of acres and hundreds of species.

8

Five Reasons to Adopt Environmental Design

Susannah Hagan

A profound and wide-ranging reappraisal of material culture, initially hijacked by geeks and hippies, is being developed within the disciplines of political science, geography, cultural theory, philosophy, economics, the fine arts, the life sciences, and—at last—architecture. Many within architecture, however, refuse engagement with this reappraisal. For them, environmentalism is embarrassing. It has no edge, no buzz, no style. It is populated by the self-righteous and the badly dressed. Its analysis is simplistic, its conviction naive, its physics dubious, and its metaphysics absurd. It is a haven for the untalented, where ethics replace aesthetics and get away with it.

If these claims were ever true, they are no longer. Other, more informed descriptions of environmentalism than the caricature above have come to predominate, descriptions in which it is as complex, demanding, and shaded as any of the intellectual obsessions that have to date inspired the avant-garde. Within architecture, environmental design is not merely a set of practical solutions to a set of practical problems; it is the complex tip of one tentacle of environmentalism. Environmentalism itself is a modernist metanarrative put through the postmodern wringer: its aims are universal, but its means are responsive to, in fact dependent upon, individual conditions. Its claim to universal validity rests on the human-as-embodied contained within a

100

physically sustaining system (nature). Since we share this condition, whatever our cultural differences, we are equally obligated to protect that which physically sustains us. This is unavoidably so, however much one might protest that nature is a cultural construct: "[I]t is not language that has a hole in its ozone layer; and the 'real' thing continues to be polluted and degraded even as we refine our deconstructive insights on the level of the signifier," writes English philosopher Kate Soper.[1] The correlate to this universal implication is the recognition that the particular is as important as the universal, and that the whole is made up of highly differentiated parts, differentiated culturally as well as materially. Our obligations, with a nod to Marx, take the form of "to each according to his or her needs, from each according to his or her use (or abuse) of nature." German sociologist Ulrich Beck sees environmentalism as a new stage of modernism, a "post-imperial" or, in his phrase, "reflexive modernism": "Modernity has . . . taken over the role of its counterpart—the tradition to be overcome, the natural constraint to be mastered."[2]

In the built environment, which contributes 50 percent of all man-made greenhouse gases, the obvious candidate for leading this "overcoming" is the architect. For although architecture's direct

MVRDV, The Dutch Pavilion for the 2000 Expo, Hannover. Copyright Rob't Hart Fotografie.

physical impact is minimal, its cultural impact is disproportionately significant, inside and increasingly outside the building professions. In fact, there are at least five reasons why schools and practices should pick up the environmental gauntlet.

The Intellectual Reason

The idea that the operations and organizations of nature are vastly superior to those of our material culture is not held just by environmentalists or environmental architects.

After the rigors of the Modern Movement, during which nature-as-model was regarded with great ambivalence, we have returned, on the basis of deeper understanding, to new kinds of nature worship. These can be found in very unlikely places: in the temples of the most arcane cultural commentary and the highest of high architecture. This is Sanford Kwinter on the seductions of matter: "Semiotic structures are binary, hierarchical, closed. . . . Matter is literally riddled with properties, dissymmetries, inhomogeneities, singularities. . . . Matter is, in short, active, dynamic and creative."[3] It is the particularity and the dynamism of nature—its capacities to repeat without repeating and to evolve, capacities that have evaded mechanical mass production—that are now within our ability to imitate in material culture (most dramatically in biotechnology). The closer these come within our reach, the more impatient certain architects get with architecture's conventional boundaries: "The exact, proportional, fixed, and static geometries, seemingly natural to architecture, are incapable of describing corporeal matter and its undecidable effects. . . . [R]ather than violating the inadequate stasis of exact geometries, . . . architecture must begin with an adequate description of amorphous matter through *anexact yet rigorous geometries*."[4] Within this energetic thinking, the goal seems exclusively formal. As far as it goes, one cannot fault it: if architecture concerns representation and if what is to be represented is shifting, then getting architecture to shift its "description" makes eminent sense. Some, of course, have already addressed this. In the 1980s, for example, Peter Eisenman was already experimenting with using geological phenomena to generate forms.[5]

What is new now is the desire to represent formally the dynamic aspects of nature, rather than the more easily represented static ones. Which begs the question: if there is such a keen interest in the way computer and life sciences are revealing hitherto unreadable workings of nature, why is the *sight* of them enough? Why isn't there impatience not only with the way architecture is representing an incomplete picture but also with the way it is *enacting* an incomplete picture? The sciences that are enabling us to see and the industries dependent on them are once again racing ahead of architecture to meld ever more deeply and inextricably with natural processes. The only architectural practice even trying to keep up is environmental design.

The Practical Reason

An enormous gulf exists between those who look to the new model of nature as the source of new generative strategies and/or forms and those who look to it as the source of new ways of constructing and running buildings. The intellectual pyrotechnics of the former are missing in the latter. The intellectual consistency of the latter is missing in the former. Reading Thomas Herzog or Glenn Murcutt is not remotely as stimulating as reading Kwinter or Greg Lynn, but reading their buildings can be. The provocation of environmental practitioners lies in what their buildings are doing, not in their discussions of it. The form of a building, however, is crucial to its environmental performance, as are its orientation and materials. At a more detailed level, the fixtures and fittings of environmental design are also crucial: photovoltaic and solar panels (to include or exclude), (day)light shelves, (day)light tubes, architectural shading devices, buffer zones, planting, hybrid HVAC systems—part mechanical, part passive—ventilation chimneys, and so on.

This environmental vocabulary is handled with great panache by some architects (for example, the British Edward Cullinan, and Feilden, Clegg, and Bradley; the Italian Mario Cucinella; and the Malaysians Hamzah and Yeang) and with little panache by too many. In architectural terms, the most interesting practitioners are the ones who break out of any kind of environmental functionalism, architects like the Dutch Mecanoo; the Anglo-German Sauerbruch and Hutton, and Alsop and Stormer; and the French Jourda, Perraudin, and Edouard François (of interest for his imagery rather than his energy efficiency). The range of strategies and architectures is enormous, between firms and sometimes within firms, as they respond to different cultural and environmental contexts. Some designers follow environmental

thinking to its logical conclusions; others do not; some do it sometimes, and sometimes not.

The difference between those architects who have gone only a
little way and those who have gone far down this path is instructive.
Perhaps the most prominent architects to have taken up a revision
of Modernism are those once labeled High Tech and rechristened
Eco-Tech by Catherine Slessor: Richard Rogers, Norman Foster, and
Nicholas Grimshaw.[6]

Mecanoo, Faculty of Economics and Management, Utrecht, 1995. Copyright Christian Richters.

They have a clearly stated intention to improve the environmental performance of their buildings and an unspoken desire not to stray too far from their previous practice and its commercial success. For this reason, they will go so far and no farther with a re-formed Modernism. For example, their materials are still high-performance, industrially produced metals and glass with a great deal of "embodied energy."[7] Grimshaw's Börse, Rogers's DaimlerChrysler headquarters, both in Berlin, and Foster's Swiss Reinsurance Bank in London are cases in point.

Positions evolve, however. Rogers's Tribunal de Grande Instance in Bordeaux (1998) extends the firm's usual palette of materials with wood. Each of the seven freestanding law courts has a glulam (laminated wood) superstructure on a concrete base, clad with cedar strips on the exterior and plywood on the interior. Wood is a good material environmentally, not only because it resists conducting heat and cold, but also because it adds no carbon to the atmosphere (except in its transportation and cutting). This makes Rogers's specification of aluminum in the Bordeaux courts' office block and copper for the roof inconsistent—but it takes time for internal contradictions within a practice to become intolerable.

Trade-offs will probably always be made between fossil fuel consumption, architectural effect, and structural performance. The real contribution of Eco-Tech has nothing to do with maximizing energy efficiency or methodological consistency. Its value lies elsewhere: first, in pushing the technological envelope of environmental design— hardly surprising, since innovation is its raison d'être—and second, in acting as a Trojan horse. These architects' very reluctance to allow their architectural identity to be diluted by environmental design has meant that they have been able to hold on to powerful commercial and institutional clients who are reassured by the familiarity of the house style and yet must at least pay lip service to environmentalism. This is by no means adequate, but again, it takes time for contradictions between action and intention to become embarrassing enough to resolve, and Rogers, at least, goes considerably farther than environmental window dressing in his work.

In contrast, Rogers's former partner Renzo Piano has resolved more of the contradictions between High Tech and environmental practice by leaving High Tech's visibility behind. An example of this shift is the firm's Tjibaou Cultural Center for the Kanak population of New Caledonia. The governing ideas for both the site and the Center were derived from Kanak mythology, which is in large part derived from climate and topography. Around the Center winds the Kanak Path, which represents the four stages of Kanak culture creation—agriculture, habitat, the country of the dead, and the spirit world—each of which is closely associated with particular stones, plants, and trees. The Center itself is arranged as a path that reproduces the organizational idea of the Kanaks' ceremonial path, which is lined with trees and ends in the chief's hut. Instead of trees, programmatic functions line the Center's ceremonial path, enclosed in the ten wood "cases," or huts, that make up the "village."

The cases that dominate the design of the Tjibaou Cultural Center were initially conceived for cultural, not climatic, reasons—they refer to the Kanaks' own huts—but they were modified to perform their environmental function more efficiently. The cases thus evolved from a conelike shape that echoed the conical roofs of the Kanak huts to more of a cone cut in half, to increase air flow for ventilation. In both the traditional vernacular and the contemporary approach, the strategy is climatic, taking advantage of the Pacific trade winds. The cases are made of iroko wood, with laminated wood elements up to twenty-eight meters high supporting horizontal curved slats that allow free air circulation between themselves and the louvered inner skin. The louvers are computer operated, designed to open automatically to their full extent when there is a gentle breeze and to start closing when wind speed increases. If the wind shifts direction, ventilation is through the much lower front of the building, evacuating through the top of the double skin. The design was developed through wind tunnel testing and computer simulations carried out by Ove Arup and Partners and the Centre Scientifique et Technique du Bâtiment.

The result is a design that may remind the Kanaks of their own minimal built culture but in no way seeks to imitate it. Piano was adamant about avoiding the slightest hint of kitsch. Instead, he imitates the Kanaks' own interactive response to climate. This is done through a similar attention to passive ventilation but with very different materials and at a very different level of technical complexity. The laminated wood pillars, for example, are set into a cast steel foundation, which had to be transported across the Pacific to the site, as did the computers and the louvers. Although to construct a building of any size and complexity on an unendowed island would always require imports, for environmental fundamentalists the choices would have been different. Doubtless greater energy savings would have been achieved, but almost certainly not to such noticeable effect. What is striking is the way Piano has allowed climate to inspire form to a degree his former High Tech comrades have not, with the result that his architecture has become diverse in a way theirs has not. This is a choice. Some of the first-generation Modernists—Aalto, Barragán, and, off and on, Le Corbusier—chose to go the other way, remaining within the idiom of the International Style while responding architecturally to climate—with brise soleil, for example.

The Technical Reason

Environmental design is demanding. It is not merely a question of a few simple, intuitive moves like shading a south-facing Northern Hemisphere facade in summer, and protecting a north-facing Northern Hemisphere facade in winter, though it is astonishing how even these actions are beyond many. Environmental design began as building physics, and that is where it ends. The more complex a space and program, the more complex the physical interactions taking place both within it and between the interior and the exterior. The materials of the building are also of central importance. Concrete-as-playdough will not do. Materials have very different conductivity and reflectivity properties that directly affect environmental performance. The development of powerful environmental software is being driven by the need to test the complex physical behavior of a passive or hybrid building before it is built to see whether the control of the internal environment is still satisfactory after the reduction or renunciation of conventional energy-expensive mechanical systems. The more ambitious the design, the more this modeling is necessary, since opening a building up to the outside again results in many more variables than a sealed building has to address—fluctuations in solar radiation, wind velocity, humidity, and so on. At the same time, the environmentally designed building must meet much higher comfort and performance levels than vernacular architecture, the model for a less energy-hungry architecture.

The need to understand the patterns of interaction between the forces of nature and buildings has produced a demand for computing power large enough to perform the analysis. It is only through our ability to import the real complexities of the natural world into cyberspace that we can test the environmental performance of not-yet-constructed buildings. Programs like Radiance and Ecotect, made for designers, not engineers, are as stimulating as they are demanding. Radiance is a "physically based lighting simulation program" that demonstrates the quality and intensity of daylight in any part of any building over a day or a year. Ecotect is a "3-D modeling interface fully integrated with acoustic, thermal, lighting, solar, and cost functions" that supports conceptual as well as final design. The point is that environmental design does not rule out complexity of form or program. A repositioning of material culture opens up new practices as it shuts down others.

The Economic Reason

In Europe, environmental reform comes from the top down. The European Union, through a mixture of legislation, directives, economic incentives, and subsidies, is attempting to redirect its vast bulk toward a less damaging relationship with the environment, built and natural. The United States, under the current administration at least, is resisting any such intervention tooth and nail. Instead, the engine for change is the conventional villain of the piece: big business. This is not in obedience to any moral imperative but from its usual regard for the bottom line.[8] "Industrial ecology" can save companies millions of dollars by treating industrial processes as if they were biological and by transforming production and consumption from a linear entropic process to a circular energy-efficient one.

There is no waste in nature because all its "manufacturing processes" are interrelated through all scales from the local pond to the globe. What one organism no longer needs is used by another. The biosphere was constructed from these relationships, each further level of complexity emerging from a symbiotic relationship with the levels below. Industrial ecology imitates this interrelatedness. Waste becomes another sellable or exchangeable commodity. Instead of our producing, say, steel, and its waste being dumped as an unwanted by-product, that waste is used by another enterprise for another industrial process. (For instance, blast furnace slag from steel manufacture can be used as a cement replacement in concrete. In Japan they use up to 80 percent blast furnace cement to 20 percent ordinary portland cement. In Europe, structural steel is often recycled.) Making the waste of one production cycle the raw material of another has yet to become commonplace, but where it is being tried, it is bearing fruit, as in Kalundborg, Denmark, where:

> The project encompasses an electric power plant, . . . an oil re-finery, a pharmaceutical plant, a wallboard factory, a sulfuric acid producer, cement manufacturers, local agriculture . . . and nearby houses. In the early 1980's, [the electric power plant] started sup-plying excess steam to the refinery and pharmaceutical plant. It also began supplying waste heat for a district heating system, allowing 3,500 oil furnaces to be shut off. In 1991, the refinery began remov-ing sulfur for its gas, selling it to a sulfuric acid producer. . . . [The electric power plant] is now selling its fly ash to the cement manu-facturer and will . . . sell waste gypsum to the wallboard plant . . . and the pharmaceutical plant is turning its sludge into fertilizer for local farms.[9]

There is no reason, given the diversity of environmental practice, that it cannot be diversified even further by those who have yet to engage with it. In Europe, remaining aloof is less and less an option, not only because energy efficiency requirements are becoming more de-manding, but also because firms are competing in a market in which those with this expertise have a competitive edge over those with-out it. Not that everyone in Europe is rushing to upgrade his or her practice. It takes time and money, and many firms would rather wait until they are pushed. Paul Hyett, president of the Royal Institute of British Architects, recently complained at the inaugural meeting of the Global Alliance for Building Sustainability about the slowness of the profession in the United Kingdom to incorporate this new prac-tice: "I can't force on my members obligations that would make them uncompetitive. . . . [Nevertheless,] is it necessary to burn fossil fuels for ventilation and lighting every time someone uses a toilet? . . . Buildings are unintelligent and their design is unintelligent."[10]

Although it may not be moving quickly enough, the profession in the European Union is moving, if for no other reason than clients are beginning to see that although a low-energy building may require a larger initial capital outlay, that outlay is "paid back" in dramati-cally lower running costs. Whether moving into their own buildings or trying to get others to move into their buildings, clients see this as increasingly important. Oil will not always be cheap in the United

States, and clients will begin to respond in the same way they do in fuel-expensive Europe.

The Pedagogical Reason

If, in answer to accusations of foot-dragging, architecture offices can plead practical difficulties—which should be addressed with practical help—architecture schools have no such excuse. Their remit is to prepare the next generation of architectural thinkers and doers. They may protest, with good reason, that they are not able to find sufficient numbers qualified to teach environmental design, but this is, at least in part, a problem of their own making. Environmental design has been around for thirty years. Given the usual degree of inertia, that is twenty years of training lost, twenty years worth of skilled teachers and practitioners unavailable, and therefore a much smaller pool of expertise. The failure to address the issue sooner also means that the pedagogy remains in its infancy. How does one teach architects something that is a technical practice, but a technical practice like no other, one that inhabits the building process from conception to construction and carries enormous formal implications? What is the best way of teaching architecture students to juggle the formal and the environmental so that neither is overcome by the other?

At the Architectural Association (AA), for example, where I taught, there are both long-standing and emerging teaching practices from which to learn. Neither is entirely satisfactory, but each points to ways a more effective synthesis may be achieved. The long-standing practice, the postgraduate Environment and Energy Program, is found in the AA's graduate school and has been running for seventeen years, a distant moon orbiting the boiling planet that is the rest of the school. Young architects, some newly qualified, some with a few years in practice, spend an intensive year acquiring the principles of environmental design, an ability to think their way around the subject, and, above all, a proficiency in the most recent environmental software. Computer analysis and simulation dominate the course. This is in some ways good—it is rigorous and concrete, and enables students to see clearly which decisions are environmentally advantageous—and in some ways unsuccessful: students do not have time to integrate the new practice into their established ways of designing.

The alternative pedagogical approach is in the main school of the

AA, at both degree and diploma levels. Here, certain design units are deciding, quite independently of the school and each other, to incorporate some form of environmental design into their syllabus, usually as one of several simultaneous means of generating form. No experts are involved in the week-to-week teaching—it is strictly bottom-up—but the pedagogical model that is evolving is, for architects, potentially more promising: students develop conceptual frameworks, which often have nothing whatever to do with "the environment," and simultaneously research aspects of environmental design on a need-to-know basis. Their intellectual inquiry is therefore both cultural and environmental, and the dialogue between the two generates the final design. The outcome, however, is limited by the nature of the input. Students may successfully research photovoltaics, reed-bed water purification, and recycled materials, but they are not able to test either their empirical observations or their final decisions. They do not have the ability, so painstakingly acquired in the master's program, to analyze or simulate. What is needed, beginning at degree (undergraduate) level and continuing at diploma (graduate) level, is a combination of the Environment and Energy and the design unit approaches.

This pedagogical field, like its practice, is open for innovation and experimentation. In Europe, demand is growing among students for more environmental course content, a desire firmly backed by the Royal Institute of British Architects, which now requires all schools to incorporate sustainability into their core curricula. Recruitment is beginning to be affected: students who do not find what they are looking for in one school choose another. Students already in a particular school are increasingly critical of tutors who cannot or will not "deliver" environmentally. In fact there is a case to be made for a massive retraining of teachers. It is hardly their fault the environment was not a focus of their education, but their lack of knowledge—practical and conceptual—is becoming a threat to their school's success.

For those impervious to moral imperatives, made all the more resistible by the smug sanctimoniousness of the converted, surely there is enough intellectual and aesthetic challenge within environmentalism to merit a few toes in the water. The concerns of the current avant-garde and those of environmental architects meet in nature, and the refusal on both sides to acknowledge this common ground is as obsolete as it is limiting. The idea, prevalent among the unenthusiastic, that the exigencies of environmental design pose an ominous threat to their creative freedom is unfounded. Within the framework of a relationship with the biosphere a little more mature than "gimme, gimme," certain limits are now necessary. There are, or need to be, limits to certain forms of material exploitation, but within those parameters, we find ourselves in a condition described by University of Essex sociologist Ted Benton as "bounded but unlimited."[11] Few architects have even begun to explore the implications of the new embedded within limit—a limit that is material, not intellectual—hampered as they are by the association of the new with the limitless. Why environmental design should be perceived as having a more disastrous effect on creativity than any of the other limitations architects are faced with—of budget, client demands, building regulations—remains a mystery.

On the contrary, limits can serve as a means of making design decisions less arbitrary, more grounded in the "real." Perhaps this is one of the causes of resistance to environmentalism—not practical worries about the time and money required to get up to speed, but a sometimes explicit, sometimes subliminal resistance to architecture-as-matter. But in a culture increasingly capable of merging nature and culture, why on earth are thoughtful talented people still addressing only one end of an enormous range of new possibilities?

2003

Notes

1. Kate Soper, *What Is Nature?* (Oxford: Blackwell, 1995), 151.

2. Ulrich Beck, *Risk Society* (London: Sage Publications, 1992), 185.

3. Sanford Kwinter, "The Genius of Matter: Eisenman's Cincinnati Project," in *Re-Working Eisenman* (London: Academy Group Ltd., 1993), 93.

4. Greg Lynn, *Folds, Bodies, and Blobs* (Brussels: Books-By-Architects, 1998), 83.

5. Peter Eisenman, "Centre for the Arts, Emory University, Atlanta," in *Folding in Architecture: Architectural Design Profile No. 102*, ed. Greg Lynn (London: Academy Group Ltd., 1993).

6. Catherine Slessor, *Eco-Tech: Sustainable Architecture and High Technology* (London: Thames and Hudson, 1998).

7. "Embodied energy" is the energy (almost always fossil fuel energy) required to extract, process, and transport the materials that make up a building. The aim in environmental design is to reduce the embodied energy of a building as much as possible. That said, the difficulty of systematizing such reductions is such that embodied energy is rarely, if ever, taken into ac-

count when judging the energy efficiency of a building, not even in building regulations in the United Kingdom.

8. "In the 2001 report by Baxter International, an Illinois medical products maker, the company detailed how reductions in energy and water use and improved waste disposal and recycling over the past seven years cut costs by $53 million this year. That savings amounted to nearly 10% of its net income." *Time,* August 26, 2002, 30.

9. Sim van der Ryn and Stuart Cowan, *Ecological Design* (Washington, D.C.: Island Press, 1996), 114.

10. "Hyett Addresses Building Summit," *RIBA Journal,* September 2002, 97.

11. Ted Benton, *Natural Relations: Ecology, Animal Rights, and Social Justice* (London: Verso, 1993), 177.

9

Invitation to the Dance: Sustainability and the Expanded Realm of Design

Peter Buchanan

Complacency in the face of the abundant evidence, from the most reputable of sources, about the gravity of the burgeoning environmental crisis is wanton hubris. After all, the very first civilization, Mesopotamia, was brought down by environmental degradation. The once Fertile Crescent withered to desert as salts leached by irrigation sterilized its fields, and deforestation of a hinterland reaching the cedars of Lebanon reduced rainfall.[1] Since then, similar factors contributed to the decline of the great European empires of Greece, Rome, and Spain, as well as some in Central America.[2]

Each time, however, environmental degradation—Greece's soil dislodged by overgrazing and olive roots, Rome's North African breadbasket turned into the Sahara, Spain's rainfall reduced by the felling of forests for shipbuilding—and the collapse of the dependent culture were relatively contained geographically. Other lands and emergent cultures succeeded these on our ever-evolving earth. Indeed, limited ecological collapse could be seen as evolution's way of terminating sclerotic cultures to replace them with dynamic, still-evolving ones. But now, as the world's cultures coalesce with globalization, it is the whole biosphere and all its peoples that are severely threatened, a situation perhaps again applying pressure for further evolution.

The current crisis results from privileging the rights and well-being of one species above those of all others. Even massive immediate action would not save from extinction a large proportion of nature's species and thus prevent compromising the biodiversity on which the planet's ecological health depends.[3] Also privileged above humankind's myriad extant cultures is the dominant form of consumerist corporate capitalism being spread by globalization. Although it, and the scientific-industrial civilization it is evolving from, have brought many benefits (material well-being, medical care, human and gender rights, knowledge and education), this dominant culture is unsustainable, for many reasons. Not least, its rapacious appetite for biologically generated resources and prodigiously wasteful production and distribution processes extracts these resources quicker than the earth can replenish them and dumps toxic wastes faster than they can be neutralized and absorbed. This situation will worsen as more of the planet's growing population aspires to the still increasingly profligate lifestyles of the developed world.

Of the many symptoms of overstraining the earth's capacities, the most conspicuous and pervasive are global warming and climate change. These already cause millions to suffer from drought and desertification, from flooding and other forms of violent weather, and in hunger, homelessness, and disease. Moreover, studies suggest that the window of opportunity for changes radical and widespread enough to save the rest of us from similar suffering is closing fast. Even if we take that opportunity, the legacy of pollution will continue through generations affected by such scourges as hormonal disruption and chromosomal damage. Surely the most urgent challenge of our time is to bring about a sustainable global civilization. This would reverse the anthropocentric and homogenizing tendencies of scientific industrialism to engender respect for all forms of life and allow a myriad of locally rooted cultures to thrive. It would not be regression but evolutionary progression, not least in encouraging each culture to evolve in its own way to create ever-higher levels of diversity and order—the basic mandate of evolution and criterion of ecological viability.

Alarmingly, the term *sustainability* is itself regularly misappropriated and abused. Peppered meaninglessly through corporate, academic, planning, and architectural prose to make proposals palatable, it triggers cynicism instead of promising the salvation of much of what we cherish, if not yet of our species' very survival. If the term could be redeemed, sustainability would denote long-term viability, not some tiny advance toward that ideal, as with most of what "greenwash" currently labels sustainable. Compounding confusions, sustainability is used in a narrow as well as in a correct, broad sense. Used narrowly, it emphasizes only environmental and ecological concerns as they manifest both locally (in pollution and the losses of species, biodiversity, natural habitat, and beauty) and globally (in resource depletion and climate change). Hugely important as these are, addressing them alone could never deliver sustainability. Sustainability, in its broad, full sense, also requires economic viability and greater social equity. Without the former, any environmental initiative is bound to fail; without reversing the trend to increased inequity between and within nations, political and social disaffection and instability (whose consequences include terrorism) remain inevitable, and long-term viability improbable.

Moving toward equity is seemingly the most insurmountable problem, a dimension of which is dramatically highlighted by ecological "footprint" analysis.[4] All earth's productive land and sea, divided equally among its population of six billion, leaves only 1.9 hectares (4.7 acres) to support each person. A Bangladeshi's subsistence needs only half a hectare (1.2 acres) to support it, but the average American's lifestyle demands 12.5 hectares (30.9 acres).[5] This suggests that for the whole world to enjoy America's standard of living would require another five earths. In fact, if America used more renewable resources (particularly for energy), recycled more, and introduced the far more efficient means of production and distribution already available, or readily developed, its footprint could be reduced to a fraction of its present size without compromise to its standard of living.[6]

Ultimately, the great challenge is to lessen the developed world's ecological footprint while offering the developing world a more sane and seductive model to emulate. This requires shifting from the goal of a standard of living to that of quality of life, transforming the drive to simply get and consume into the profoundly different one of pursuing deep psychic fulfillment—a step forwards, not backwards, as it is too often portrayed. Achieving sustainability, then, requires attention to psychology and even spiritual issues, to satisfy values deeper than advertising-induced desire, needs like having felt connections to nature and communities.[7] Significantly, this is what is promised from comprehending, internalizing, and living in accord with a worldview that ecology and evolutionary, chaos, and complexity theories have long been shifting to: the universe is not a dead clockwork mechanism but a living process, constantly unfolding and creative. The profound psychological impact of this shift is that we need no longer feel alienated from the world, nor compelled to defend against this feeling through acquisitive consumption, but can instead disencumber ourselves to open up to and feel an integral part of this astounding and benevolent planet.

It is hardly surprising, then, that some herald the quest for sustainability as inevitably prompting an epochal cultural transformation. This will have to encompass the reappraisal and redesign of everything from modes of agricultural and industrial production and distribution, to settlement patterns and transportation, legal and economic systems, taxation and education, lifestyles, social customs, and even redefining what it means to be fully human. This immense and exciting challenge will color every aspect of the foreseeable future and require not only great political intelligence and will but also the participation of all sensitive, creative, and responsible people. Because of the huge global population to be supported, and its high aspirations, this cannot be achieved with any recidivist or Luddite moves. Instead it will have to offer the enticing improvements without which change would not be readily endorsed and embraced, changes like the enriched social dimension brought to buildings such as office blocks and hospitals by energy-conserving atria with their opportunities for casual encounters on café terraces and at kiosks in verdant settings. In particular, it must accord with newly emergent (or reemergent) values in which our deepest satisfactions and destiny are to be found in living in harmony and intimate contact with other people and nature. The result should be a creative renaissance, especially in the arts, which might contribute by offering inspiring visions of possible futures, evoking longings for living in accord with the planet's rhythms and regenerative capacities.

To be achieved, sustainability needs to be pursued at all levels, from the governmental and the corporate to the personal—through international accords and government diktats, if necessary. Europe—the European Union and the north European countries of Scandinavia, Germany, the Netherlands, and the United Kingdom, in particular—is seriously committed to sustainability; these countries have all made good progress in matters like cutting greenhouse gas emissions far beyond the proposals of the Kyoto accords, scuppered by the Bush administration. (Some reasons for the differences between Europe and the United States are elaborated in endnote eight.)[8] New directives and legislation continuously ratchet up performance standards for recycling and air quality, reducing waste and energy consumption, and increasing renewable energy supplies. The European Union has funded research by architects designing energy efficient buildings and pays for test installations of as yet unproven and uneconomical technologies.[9] (This speeds testing of promising technologies so that those proven viable will be adopted on a scale that will quickly make them economical.) The current U.S. administration is virtually alone among developed nations in its recalcitrant obstruction of moves toward sustainability.[10] Eventually this must change: under pressure from the electorate and the inspiration of foreign example; as more American manufacturers discover the enhanced profitability and export advantages of green products and production; and as the insurance industry baulks at the enormous costs of global warming–induced hurricanes, flooding, and fires.

The environmental design disciplines—architecture along with urban design, regional planning, and landscape architecture—will inevitably play a key role in the quest for sustainability. It is, after all, an *environmental* crisis that looms, and the design of the current built environment contributes immensely to this crisis, in its wastefulness of land, energy, and commuting time and in the lifestyles it facilitates. It also constrains how much we can change these lifestyles. But inventive design can be an effective agent for positive change, especially if applied to both the built environment and to planning strategies and legislation. No matter what coordinated design can achieve, we must be realistic and recognize that architecture alone cannot deliver sustainability. The term *sustainable architecture* is therefore problematic—as is *green architecture,* because some wrong-headedly still see it as implying a single consistent approach or aesthetic. The looser term *green buildings,* or even the cumbersome *environmentally responsible buildings,* is to be preferred because green designs need to be as varied as the places and cultures in which they are located.

Buildings account for nearly half the energy consumed by developed countries. Most of this energy powers the artificial light and air-conditioning that contemporary nonresidential buildings use even on bright and balmy days. Generating this electricity is the single greatest source of greenhouse gas emissions—although transport, particularly road vehicles, especially in the United States (exacerbated by the fad for four-by-fours and SUVs), contributes an ever-increasing proportion. The sprawl of America's cities and suburbs causes huge wastes of energy and (commuting) time, and of land, paved over or turned into the scrubby residual space along roads and around buildings. America's commercial vernacular of shoddy glass or windowless boxes afloat in a nowhere of asphalt and nondescript landscaping annihilates any sense of place, community, or connection with nature; these buildings exemplify and entrench the deadened sensibilities that have tolerated the destruction of the natural and civic environment. The continuously artificially lit and ventilated interiors of these energy guzzlers, hermetically sealed behind sound-deadening tinted glass, poison the soul but also the body with their off-gassed fumes and airborne abraded particles.

Such buildings mark a cultural nadir, and going forward will, in part, have to resurrect lessons from earlier ages and other cultures. Green design is not new: vernacular architectures are, without exception, highly resource efficient, using local materials to make buildings well adapted to local climate and ecology. Today's green designers seek inspiration from them, as did early Modernism, which grew from proto-green roots such as the Arts and Crafts and Art Nouveau movements, as evidenced in the writings and designs of Frank Lloyd Wright and even Le Corbusier. But emerging green design pursues a broader set of intentions more self-consciously and rigorously than ever. Its widening range of ecological concerns can be subsumed under four overlapping goals taken up over the years in this order: conservation of local landscapes and ecology, then the reduction of, in very rough sequence, energy consumption and greenhouse gas emissions, waste and forms of pollution other than greenhouse gases, and, now, ecological footprints.

Among the many critical reactions to mainstream culture in the 1960s was dismay at the destruction of natural beauty and of valuable yet vulnerable ecosystems by thoughtlessly rapacious developers, as well as at air and water pollution manifesting as smog, acid rain, and dead rivers and lakes. This resulted in conservation movements and pressure for architects and landscape architects to protect and enhance "natural capital" (as it is now called)—incalculably precious resources accorded no value in economics. Design, it was recognized, should thus start with surveying ecosystems, microclimates, geology, and hydrology and understanding their dynamic interdependencies. The design scheme should then be worked around which of these the surveys reveal to be most precious for ecological health, beauty, or rarity and should enhance rather than impinge upon them.

Influencing landscape design more than architecture, this approach is exemplified in books like Ian McHarg's classic *Design with Nature* (1969) and schemes like Lawrence Halprin's for Sea Ranch in Big Sur (1967).

Then in the 1970s came the Arab-Israeli wars and oil shortages and price hikes. These led to the sporadic investigation, largely in domestic projects by the American counterculture, of energy-efficient design and the harvesting of renewable energy with such crudely simple, yet relatively effective, devices as solar heat–absorbing rooftop collectors and Trombe walls. Through the 1980s, growing awareness of global warming and the anticipated ravages of climate change gave new impetus to investigations of green building in Europe, rather than in the United States, where the financial and political power of the oil industry suppressed acknowledgement of these realities, and its cheap oil eliminated any economic incentive.

To diminish and eventually cease contributing to global warming (whose effects will persist for many centuries even once greenhouse gas emissions have returned to pre-industrial levels) require reducing energy consumption and switching to renewable energy sources. Better insulation and the use of high-efficiency fixtures and mechanical plants reduce energy consumption in all buildings. New buildings can reduce energy consumption much further by maximizing natural light and ventilation, trapping and storing solar heat where and when beneficial, and excluding it where and when not. With high insulation standards, interiors stuffed with electronic equipment, and efficient heat exchangers recycling even the heat emitted by lights and occupants, contemporary buildings require little heating. Instead, they need cooling, the most energy-hungry component of conventional air-conditioning. Green buildings use various low-energy cooling methods such as purging heat with cool night air, and chilling the structure or ceiling panels by pumping groundwater through them or water chilled by solar-powered heat pumps. Green buildings are also often wrapped in a secondary glass skin, outside of the weather-excluding membrane. This membrane lowers energy consumption by trapping a winter jacket of warm air, can be opened up to let breezes through on temperate days, and on still days creates "stack effect" updraft breezes. Only on the hottest days is the inner skin sealed and the building ventilated mechanically.

Foster & Partners' Commerzbank tower in Frankfurt (with structural engineers Ove Arup & Partners and mechanical engineers

Foster & Partners, Commerzbank, Frankfurt, Germany, 1997. Copyright Dennis Gilbert/VIEW.

J. Roger Preston & Partners), one of the first of a handful of green sky-scrapers in Germany, demonstrates that the devices delivering energy efficiency can also bring the improved quality of life intrinsic to the green agenda. Its offices are naturally lit and ventilated, with backup systems of cooled mechanical ventilation and heating required only in extremes of summer and winter. Perimeter offices have opening windows behind outer sheets of glass that intercept rain and wind while admitting air. Inner offices look across the hollow core of the triangular tower and out to the city across sky gardens. These spiral up the tower, letting light and air into the core and inner offices, and providing a verdant setting for informal meetings and coffee breaks. They are part of a hierarchy of social spaces that range from the meeting spaces on every office floor to the glass-roofed ground-floor "plaza" with restaurants, where employees may mingle with colleagues from the older bank tower adjacent and with the general public. Having all these social facilities, as well as personal control of comfort conditions,[11] the building is hugely popular with clients and employees.[12]

Renewable energy can be harnessed from several sources. The sunlight, heat, and wind that constantly impact a building contain many times the energy required for its lights and equipment. Harnessing this ambient energy could eliminate the huge losses of transmitting electricity from distant wind or wave farms. But photovoltaic cells require too long a payback time to be considered commercially competitive, and wind turbines as intrinsic parts of a building have not been fully perfected. Yet within a decade or less they will be viable, especially if used in conjunction with clean fuel cell technology, which will always depend on other forms of energy to charge the cells. In the short-term, the most viable and economic way to generate electricity from renewable sources is with local combined heat and power (CHP) units that consume biomass (vegetable matter) that releases only the carbon dioxide it had withdrawn from the atmosphere.

That efficient use of renewable energy can also result in resonantly symbolic design is shown by the renovation of the Reichstag, also by Foster & Partners,[13] which has its own CHP (currently running on rapeseed oil) and stores excess hot and chilled water in caverns deep underground. Subjected to arson by the Nazis and stormed by the Red Army during the invasion of Berlin, the building is steeped in tragic history. Now its original bombast, so redolent of the kaiser's imperium, has been transformed by a new, transparent and light-filled core, the assembly chamber, explicitly suggesting open and democratic government. Crowning the chamber and conspicuous outside is a new steel and glass dome rising as lightly and optimistically as the original (also steel and glass) had borne heavily down. Hanging from the center of the new dome is a cone, which dominates the chamber as an explicit symbol of Germany's commitment to the green agenda. This draws warm stale air out of the chamber while its mirrored facets reflect daylight down into it or, when parliament sits late, project shafts of artificial light across the night sky, evoking the crystalline *Stadtkrone* proposed as symbols of communitarian democracy and peace by leading architects after Germany's defeat in the First World War.

Buildings consume energy in the extracting, transporting, manufacturing, and erecting of their materials and components. Green design seeks to minimize this "embodied energy" by making maximum use of local materials (minimizing their transport) of inherently low-embodied energy (wood being the lowest and aluminum among the highest). Using recycled materials, deploying all materials to facilitate their eventual recycling, and designing buildings to be robust and adaptable to foster their longevity further conserve embodied energy.

But even the most efficient use of energy entirely generated from renewable sources achieves little if the building is beyond the reach of public transport, cyclists, and pedestrians. Hence, the green agenda advocates reconsolidating urban settlements not only for energy efficiency but also to offer lifestyles commensurate with twenty-first-century aspirations. This involves the reuse of "brownfield" lands and various strategies to increase densities in areas already well served by public transport.[14]

Waste and pollution have been concerns since the 1960s, but they are still major problems, the solutions for which will come only when both are transformed into resources for new products and processes (wastepaper is treated to be building insulation, excess heat from one building warms its neighbor, and so on). Buildings pollute and are polluted in many ways. Greenhouse gases are emitted from generating electricity for buildings. Pure water, probably now our most precious resource, is squandered in buildings. An ideal of green design is that the water leaving a building be as pure as that entering—a seemingly remote possibility already being achieved by "living machines" and "constructed wetlands" that use plants, bacteria, fish, and snails to purify water on-site to potable standards.

The air within buildings is often dangerously polluted by off-gassed chemicals and those in airborne particles abraded from furnishings. Eliminating these toxins from both products and their production processes requires critical reappraisal and redesign of their whole lifespan from extraction, through manufacture and use, to recycling and final decay. These products—whether building components, fittings, or furniture—will also be designed so that they, parts of them, or their materials can easily be recovered and recycled, rather than merely "downcycled."[15] These relatively new procedures result not only in benign products but also in far more efficient and less wasteful manufacturing processes. Here the green agenda is set to make great strides as manufacturers discover—as Scandinavians did in the 1980s and some American companies have more recently—the economies and profits in "lean" and green manufacturing.[16]

Since the 1992 Earth Summit in Rio de Janeiro, the green agenda has expanded to include the quest for full sustainability and for developed countries to curtail their ecological footprints drastically. All measures outlined so far help reduce these footprints: recycling, for instance, lowers wastes the earth must absorb and also the resources to be extracted from it. The challenge to developed countries might seem insuperable, yet on the microscale skilled design can drastically reduce the ecological footprint of new construction. This is dramatically proven by BedZED (Beddington Zero Emission Development), a dense housing-cum-business estate in south London by Bill Dunster Architects, which comes as close as seems possible for design (of more than just buildings) to offer a sustainable, high-quality way of life.

Behind BedZED's terraces of south-facing housing, warmed in part by full-height conservatories and ventilated by rotating wind cowls, are workshop-offices, the roofs of which are used as garden terraces. On site, a CHP fueled by wood chips (from trees pruned

Bill Dunster, BedZED, Surry, 2001. Courtesy Bill Dunster Architects/ZEDfactory.

by the municipality) provides electricity and hot water, and a "living machine" treats sewage. Photovoltaic cells shading the conservatories charge electric cars (the payback time on capital investment against fuel costs saved then being only fourteen years as opposed to seventy years if the photovoltaic cells powered buildings), which all residents may use. Local organic farmers, whose produce is sold on-site, collect composted organic waste. A Londoner's eco-footprint is calculated to be 8.3 hectares (20 acres),[17] as opposed to the UK average of about 7 hectares (17.3 acres). By simply moving into BedZED, a resident's footprint is reduced to about 4 hectares (9.9 acres), but if all available facilities are used (renting an on-site workplace, commuting from the nearby train station or using a bicycle and electric car, shopping at the farms stall, etc.), this can be reduced to 2.2 hectares (5.4 acres),[18] not far above the equitable ideal of 1.9. Yet building costs for BedZED were only 20 to 25 percent more than those of the volume house builders, the extra because of all the research involved and the development of new products such as the wind cowls and the heat exchangers within them, costs that will not be borne by future schemes. The results are so enticing that all dwellings sold as soon as they came on the market, with their buyers telling journalists they were willing to pay a bit more because of the quality of life (especially due to the units' sunniness and gardens) and low utility bills the project offered.

Amazingly, there are still architects, particularly Americans, who offer lame excuses and willful misunderstandings to justify not taking even the smallest steps toward green design. They claim to be powerless to effect change because architects are responsible for only a tiny proportion of what gets built, which in turn is tiny as a proportion of the existing building stock, and are constrained by such commercial realities as tight budgets and client prejudices. Besides, they say, most of the present problems, in particular architecture's contribution to global warming, will be ameliorated once more energy is generated from renewable sources and fuel cell technology comes on line. All these claims contain some truth, as well as obfuscation, but are ultimately insubstantial excuses.

Architects exercise enormous influence over the built environment, including the buildings erected without them. They introduce innovations in layout, technology, and formal fashions, and set standards and invent prototypical solutions, many of which the more general market is compelled to adopt. Everything built today follows, if often in degraded fashion, architect-designed exemplars. Architects, as individuals and as professional bodies, are also consulted on and influence the formulation of planning and building codes, and tougher codes will almost certainly be the impetus to retrofit the existing building stock to greener standards. Architects must return to dealing with the range and urgency of real issues that fired Europe's early Modernists, setting standards to be emulated by the speculative market and inspiring the public with green designs that demonstrate every possible resulting enhancement of daily life.

Because of the elimination or reduction of mechanical components, some green buildings cost no more than conventional ones.[19] But others do. Using natural light and ventilation requires proportionally more and a more elaborate external wall area, with opening windows, sunshades, and reflecting light shelves. On tall buildings especially, two or even three layers of enclosure are common. To create a building that lasts and is loved, and also because structure is usually exposed so that its thermal inertia stabilizes temperature fluctuations, construction standards are typically high. Space standards tend to be high too, allowing long-life, loose-fit adaptability. But any extra initial capital cost is quickly offset by a reduction in running costs. More significantly, in buildings in which people work, extra initial costs are recouped many times over in reduced staff costs. Postoccupancy surveys, such as of the Commerzbank, consistently show people prefer green buildings, with their healthier and more pleasant conditions, natural light, and personal control over ventilation and temperature. This preference results in reduced absenteeism and staff turnover, savings in staff training, and (though evidence is more difficult to assess and mostly anecdotal) increased productivity. In any accounting beyond the shortest of terms, green buildings prove sound investments, hence the immediate appeal to the owner-occupiers who constitute most of commercial clients to date. In time, even rental buildings will become greener to attract tenants.

Energy from renewable sources such as wind and waves is coming steadily, if slowly, on line in many parts of the world. In a decade or so, fuel cells and photovoltaic cells should be improved and cheap enough to provide clean energy for individual buildings and vehicles. (London tests its first fuel-cell buses in 2003.) These developments will diminish the contribution to global warming of deep-plan, sealed buildings and sprawling development. But these will remain unsustainable: deep-plan, sealed buildings because unpleasant and unhealthy, sprawl because it wastes land and time and suppresses community and convivial civic life. Again: sustainability is not only about curbing environmental abuse; it is even more about enjoying a saner and more just way of life.

The willful misrepresentations offered in America to justify neglect of green design are that it inevitably results in a crude and homogeneous aesthetic, is regressive, particularly technologically, and compromises the architect's creativity. True, America's green buildings have tended to be architecturally crude. But that green buildings need look similar or lack architectural quality is abundantly disproved in Europe, where its very best architects (including Pritzker Prize winners Norman Foster and Renzo Piano) are among the leading exponents of green design, producing very diverse buildings of the highest architectural quality. That green buildings are regressive is patent nonsense: such highly efficient, high-performance buildings represent the leading edge of design and engineering. With no air-conditioning plant to fall back on, there is no latitude for error in their engineering, which must draw on precision predictive modeling using up-to-the-minute computer techniques and testing in laboratories and on-site. A major reason Europe is so far ahead of America in green design is that its fee structures allow more design input, research, and testing by the engineers involved.[20]

The complaint about compromised creativity and self-expression is almost certainly the nub issue underlying the resistance to embracing green design. Green design, which cannot be achieved by merely specifying green products, is a stimulus to creativity—it throws up an ever-wider range of design issues in which every aspect of the building, its production processes, and the ways of living within have to be rethought. But, as the complainers have grasped, it also requires profound changes in working methods and a radically updated view of creativity.

Because green design calls upon skills beyond the reach of single architects, it is inevitably a collaborative venture, with experts of all sorts—including mechanical engineers and ecologists, horticulturalists and hydrologists—heavily involved as codesigners. This added stimulus again does not compromise the architect's creativity or personal signature, which depend not on working alone but on the architect's powers of synthesis and mastery of the medium.[21] Besides involving other professionals, green design collaborates with nature and its ambient energies. Creativity as the preserve of the solitary artist-architect genius imposing his will to self-expression upon the world is no longer tenable. Instead, it is about understanding the unfolding and dynamic interplay between nature and culture and treating design as if it is a process of participating in and reconciling these processes as they flower into forms that best benefit people and planet. This is the big choice we face: to move from the ego to the eco; from acting on the world to acting *with* it; from standing alone like Howard Roark to joining in the dance.

This is also sustainability's great and exciting gift to architecture: to return to it purpose and dignity as it addresses very real and urgent issues so that, after a couple of decades of wallowing by some of its most influential figures in the fashions of form and theory, it will once again inspire influence in the shaping of our environment and culture.

Green design and design methods depend entirely on the computer. (Understanding how this is so undercuts certain current architectural enthusiasms, in particular the predilection for biomorphic blobs, as a dilettantish irrelevance with little grasp of the computer's relevance and potential.) Besides underpinning the workings of our contemporary world and bringing about such new sciences as chaos and complexity theory, the computer allows us to understand climate change (through ever-more sophisticated modeling and verified predictions) and is the key tool guiding solutions to the environmental crisis.

Although houses, housing, and other small buildings can often be designed to be green merely by empirically updating existing models, the green design and functioning of larger, newer building types depend on the computer at all stages of development and function. The computer helps chart site, ecological, and climate conditions and, with a weighting factor assigned to each, enables their synthesis. Predictive computer modeling then allows the testing and refinement of alternative proposals, studying external wind flows, overshadowing, and the admission of air and sun inside, and their consequences in rates of air change and temperature fluctuations. Design at this stage proceeds not only by shaping a building but also by studying and shaping its many reciprocities with the ecology and ambient energies of its environment, processes only the number-crunching capacities of the computer can achieve and make visible. In some engineering offices today, a building might be studied in terms of its impacts on climate and comfort conditions—for instance, new wind patterns and pressures, heat and light reflections, and exhausts such as steam (and the condensation and increased temperatures this may cause)—in an area up to a mile around, both because this is consistent with the green agenda and to avoid potential litigation from neighbors. In the realized buildings, computers process the inputs of myriad sensors disposed around the interior and exterior of the building to monitor conditions and then adjust, via low-energy motors, the various openings, sunshades, and other devices by which comfort is achieved. Sophisticated software uses "fuzzy logic" and "neural networks" so that buildings learn from past performance to better predict and prepare for changing conditions of weather and use.[22]

To realize their great potential in moving us toward sustainability, the environmental design professions must profoundly change their working methods, expanding their concerns and the spatial territory and temporal period considered their domain. Architecture can no longer be about designing isolated objects, and context is more than the forms and functions of neighboring buildings. Instead design will shape, in carefully calculated ways, the processes by which buildings interact with and impact surrounding ambient energies and local and global ecologies. The design and constructional development phase must extend its concerns to cover the whole life cycle of the building, from extracting the materials and manufacturing components from them to their eventual dismantling and recycling. Design is thus a process of looking for the best solution at all stages of the building's coming into being and long life, for everybody involved and all species affected. The ultimate (but not sole) criterion of architectural judgment would be the beneficial impacts upon the health and further maturation of all the interdependent systems and species within and around the building, ranging from earthworms aerating the soil on the building's roofs and in its grounds to the residents' sensual engagement with and understandings of nature's cycles.

2003

Notes

1. Clive Ponting, *A Green History of the World: The Environment and Collapse of Great Civilizations* (London: Sinclair-Stevenson Ltd., 1991). See pages 70–72 for details about how in Mesopotamia "short-term demands outweighed any consideration of the need for long-term stability and the maintenance of a sustainable agricultural system." Much the same set of forces led to the collapse of the early civilization of the Indus valley and

the later Mesopotamian civilization centered on Baghdad, whose population collapsed from around 1.5 million at the turn of the first millennium to only 150,000 by AD 1500.

2. Ibid., 76–83.

3. Estimates of rates of extinction and the proportion of species that will become extinct vary widely. But an article in *Nature* (November 1, 2002) asserts that more than half of all plant species are now close to extinction.

4. The idea of the ecological footprint has been most fully elaborated in Mathis Wackernagel and William Rees, *Our Ecological Footprint: Reducing Human Impact on the Earth* (Gabriola Island, British Columbia: New Society Publishers, 1995).

5. *The World Wildlife Fund's Living Planet Report,* 2002 (Gland, Switzerland: The Fund, 2002).

6. See Ernst Ulrich Von Weizsacker, L. Hunter Lovins, and Amory B. Lovins, *Factor Four: Doubling Wealth, Halving Resource Use* (London: Earthscan Publications Ltd., 1997); Paul Hawken, L. Hunter Lovins, and Amory B. Lovins, *Natural Capitalism: The Next Industrial Revolution* (London: Earthscan Publications Ltd., 1999); William M. Feld, *Lean Manufacturing: Tools, Techniques, and How to Use Them* (Boca Raton, Fla.: CRC Press, 2000); and David Wann, *Deep Design: Pathways to a Livable Future* (Washington, D.C.: Island Press, 1996).

7. See my essay "Reconciling the Millennium" in *City 8,* 1997, which discusses an urban neighborhood and domestic life of the future.

8. From Peter Buchanan, *Ten Shades of Green* (New York: The Architectural League of New York, 2003): "Some European countries have had strong clerical and office worker labor unions that demanded the physical work conditions their members prefer. A greater proportion of European businesses, both large corporations and smaller companies, build and occupy their own premises (rather than those of speculative developers) and so have direct interest in pleasant work conditions, diminished running costs, and the benefits of a stable, happy, and productive work force. Much of Europe has a more temperate climate than most of the U.S., and so it is easier to ensure comfortable conditions without resorting to air-conditioning. Europe consists of relatively small countries that demand action when pollution from a neighboring country defiles their rivers or kills their lakes and forests—in contrast to what can be resigned acceptance of home-grown pollution. European architects have generally been more concerned with social and technical issues, function and performance, than American architects, whose concerns are more with form and theory. Engineers play a larger and more collaborative role in the design of buildings than in America, their proportionally larger fee affording them twice as much creative design time. In parts of Europe the banking system and funding of construction is less geared to the short term than in the U.S. and so is less inhibiting of the long-

term thinking and accounting. European building, planning, and tax codes are less likely to inhibit green buildings than those in the U.S."

9. The research on designing energy efficient buildings was organized by the European Commission Directorate General XII for Science, Research, and Development with Renewable Energies in Architecture and Design and is known as the Joule 1 and Joule 2 programs. The test installations of technologies are carried out under the auspices of the Thermie program.

10. Inertia in dealing with environmental problems stems from: short-termism in politics, planning, and profits accounting; reductive and instrumental thinking that tends to be so abstracted and narrowly focused as to miss the bigger picture; and the hubris brought by wealth, power, and ever-improving technology that assumes that these can solve anything.

11. Studies show that people prefer not only personal control of comfort conditions but also the sensual awareness of their bodies brought by gently fluctuating conditions (that open windows offer, for instance) to the homeostatic and anaesthetizing conditions once considered ideal.

12. The building's popularity is revealed by postoccupancy monitoring and the pride with which the bank opens the building to public tours on one Saturday each month. The immense social and technical success of the building is confirmed by Commerzbank's facilities manager Peter Muschelknautz, whose latest figures show that the building is performing even better than predicted, being naturally ventilated for 80 percent of the year and consuming 20 percent less electricity than expected. See my essays on the Commerzbank in "Reinventing the Skyscraper," republished in *On Foster . . . Foster On,* ed. David Jenkins (Munich: Prestel, 2000) and in *Ten Shades of Green.*

13. Foster & Partners worked with energy consultants/mechanical engineers Norbert Kaiser Bautechnik and Kuehn Bauer & Partner.

14. Despite protests from developers because of the decontamination costs frequently involved, British policy proposals are that 60 percent of all new development be on brownfield sites.

The green agenda has many other urban implications, such as preserving biodiversity and wildlife corridors, but here the focus is mainly on its architectural implications.

15. A useful distinction is now made between the ideal of recycling, in which a material is reused in its pure form, and the usual norm of down-cycling, in which the material is reused in a less pure and degraded form, still contaminated with other materials also used in the original product. See William McDonough and Michael Braungart, *Cradle to Cradle: Remaking the Way We Make Things* (New York: North Point Press, 2002).

16. See James P. Womack and Daniel T. Jones, *Lean Thinking: Banish Waste and Create Wealth in Your Corporation* (New York: Simon & Schuster, 1996).

17. As measured by the New Economics Foundation, London.

18. As measured by the architects following standard procedures.

19. A good example is the Jubilee Campus extension to Nottingham University by Michael Hopkins and Partners. A landmark in green design and one of the most sophisticated in its many coordinated systems, from the way the landscaping forms part of the climate-modifying systems to how the whole low-pressure-drop ventilation system is driven by the suction of wind-passing tracking cowls, these buildings cost only the norm for such academic buildings.

20. In much of Europe, and especially Britain, architects treat engineers as equals in creative design, rather than as subservient technicians, so that the names of many individual engineers are as familiar as those of famous architects. This tends not to be true in America.

21. Hence, the works of an architect like Renzo Piano, one of the most collaborative in his working methods and heterogeneous in his output, bear his unmistakable stamp at levels from massing to detail, or like those of Louis Kahn, who asserted his equally unmistakable imprint on the very considerable input of his collaborating engineers, Robert Le Ricolais and Auguste Komendant.

22. Besides this dependence on precision engineering and computers, green buildings of the future will also rely on biological (mainly botanical) systems for such things as shading and cleaning, cooling, and humidifying air—an area being investigated by some engineers but still at a very early stage.

10

Green World, Gray Heart? The Promise and Reality of Landscape Architecture in Sustaining Nature

Robert France

Can a few conspicuous solar homes, constructed wetlands, bike paths, recycling industries, wildlife habitat corridors, organic agricultural plots, and wind farms really be the key to saving the world? Isn't a much greater transformation needed in global economic, political, and social institutions?
—Robert L. Thayer Jr.,
Gray World, Green Heart: Technology, Nature, and Sustainable Landscape

We live in what the great American environmentalist Aldo Leopold referred to as a "world of wounds," where there is irrefutable evidence that we are balancing precariously on the brink of natural disasters: "Human beings and the natural world are on a collision course. Human activities inflict harsh and often irreversible damage on the environment and on critical resources. If not checked, many of our current practices put at serious risk the future we wish for human society."[1] This 1992 statement from a document called "Warning to Humanity" is illuminating because it does not originate from tree-hugging "green-nicks" but from more than half of living Nobel Prize winners.

The image of the earth as the Titanic moving inexorably on its collision course, the band playing and people reveling, ignorant of their imminent fate, is an often used but still compelling metaphor for our obliviousness to coming crises in nature (defined here as everything that humans have not made). It is also germane to the question: What are the realities, illusions, and efficacies of nature-sustaining design? Though champions of sustainable design may herald its role in keeping us away from icebergs like global climate change or enormous biodiversity loss, hard-headed realists have no such hope. In short, the key question is, *can* the designers who shape a small portion of our built environment offer anything more than better designed deck chairs more pleasingly arranged?

It is important to establish two caveats at the start. First, the following critique about landscape architecture pertains solely to its role, either implied or specifically stated, in fostering environmental sustainability through either realized or ostensive "green" designs. I of course recognize that this is but *one* of the many benefits accruing from the profession of landscape architecture. And second, the following discussion deals *only* with site-specific design and not with regional land-use planning. In other words, although unequivocal evidence exists that land-use planning—such as watershed management or low-impact development—makes substantive contributions to sustaining nature, the question examined here concerns the ability of landscape architects' work on individual sites to affect nature-supporting alterations that make a significant difference. As will be seen, this is not to say, however, that such spatially restricted efforts are in any way insignificant in terms of promoting environmental sustainability through both direct means of ecological restoration and indirect means of experiential education.

The Promise of Sustainable Site Design

In August 2002, a special issue of *Time—How to Save the Earth—* came out during the Johannesburg World Environment Conference. Here for the first time in the American popular press—mixed with the usual doom-and-gloom and images of people begging for food, roads clogged with automobiles, wetlands shrinking from drought, and elephants marching to extinction—were essays dealing with the role of sustainable design in moving us back from the brink of natural catastrophe. The publication marks a coming of age for a movement that ironically, while enjoying increasing popularity among the lay public, remains marginal within the design professions.[2]

But the design professions might be on the verge of a paradigm shift in their relationship to nature and sustainability. Long-time champions such as William McDonough, Amory Lovins, and 2002 Pritzker Prize winner Glenn Murcutt have been joined by a cadre of what *Time* referred to as "some of the most prominent names in architecture [who] have turned green," like, for example, Sir Norman Foster. The sentence continues, however, with the caveat that this greening by the architectural illuminati is "at least for *selected* projects" (my italics).

As long ago as 1988, the Council of Educators in Landscape Architecture charted a course by defining sustainable landscapes as those that "contribute to human well-being and at the same time are in harmony with the natural environment. They do not deplete or damage other ecosystems. While human activity will have altered native patterns, a sustainable landscape will work with native conditions in its structure and functions. Valuable resources—water, nutrients, soil, et cetera—and energy will be conserved, diversity of species will be maintained or increased."[3] Now landscape architects seem to be scrambling to embrace both the concepts and the practices of sustainable design long after this definition appeared and after a period of near silence following the publication of two solid and important books: *Design for Human Ecosystems: Landscape, Land Use, and Natural Resources* (1985), by the late John T. Lyle, professor of landscape architecture at California State Polytechnic University, Pomona; and *Gray World, Green Heart: Technology, Nature, and Sustainable Landscape* (1994), by Robert Thayer, professor of landscape architecture at the University of California, Davis.[4] Two recent books offer evidence that the profession has taken a turn. *Sustainable Landscape Construction: A Guide to Green Building Outdoors*, by *Landscape Architecture* editor William Thompson, Kim Sorvig, professor of architecture and planning at the University of New Mexico, and illustrator Craig D. Farnsworth, represents a watershed in the evolution of the education of landscape designers in sustainability.[5] Its guiding principles—keep healthy sites healthy; heal injured sites; favor living, adaptable materials; protect water; minimize paving; consider the origin and afterlife of materials; know the costs of energy over time; celebrate light, respect darkness; defend silence; and maintain to sustain—offer a set of practical alternatives to business-as-usual. And *Constructed Wetlands in the Sustainable Landscape*, by Craig Campbell, principal with Design Studios West, Denver, and Michael Ogden, president of Southwest Wetlands Group, Santa Fe, although narrower in scope, presents a unique blending of science, engineering, landscape architecture, and environmental art, together with regulatory planning and site development, to advance a vision for managing built wetlands.[6]

Academic programs are now being retooled to capitalize on the interest shown among students in sustainable design. The University of Michigan, for example, was recently seeking to hire "a designer and scholar who is knowledgeable and experienced in the application of ecological principles to the analysis and design of the landscape and built environment . . . [and who] will interact with students and faculty who have diverse interdisciplinary interests related to sustainability such as energy-and-resource-efficient building design, green structure and infrastructure, landscape ecology, healthy buildings, urban ecosystem management, and life cycle assessment."[7] And at the Harvard University Graduate School of Design, a new award— The Loeb Sustainability Prize—will soon be implemented. Available to students in all departments, the award will be given each semester for "the option studio project that most exemplifies principles of sustainability regardless of the topic of the studio." The strategy is to "raise awareness of these principles and call attention to the importance of imbedding them in the design process rather than seeing them as 'add-ons.'"[8]

The important question is, however, "How all this is being played out among practitioners, who may be out of touch with academia, with books like Thayer's, and with trend-seeking reporters from international magazines?"

More Than Greenwash?

The design professions are not immune to fads, and green design may become their new one. One can easily become cynical about the environmental realities beneath the verbal veneer of many would-be green designs. If you scratch their surfaces, you find only sustainable rhetoric. There is perhaps no more egregious example of this than "eco-revelatory design," which, as I argued in the winter/spring 2000 issue of *Harvard Design Magazine,* just tips its hat to nature while making business-as-usual look nice. This begs the question, What exactly is the "business" of landscape architecture? And does adding *green* or *sustainable* before *landscape architecture* create a redundancy or an oxymoron?

Being a landscape architect, like being an ecologist, is certainly no guarantee of being an environmentalist. The desire of designers to make a personal mark on the landscape, and of ecologists to understand the workings of nature can often be at odds with a desire to "preserve, protect, and restore environmental integrity"—the mandate of the 1972 U.S. Clean Water Act. Even many of the subset of landscape architects who profess to engage in sustainable design, though they speak lofty, self-important words about making a "green world," seem to possess gray hearts, or certainly hearts no greener than those of the environmental engineers they are quick to criticize. Motivated in 1993 by fear that "the future of the profession is at stake," the trustees of the American Society of Landscape Architects (ASLA) adopted a Declaration on Environment and Development, an attempt to encourage landscape architects to play a "key role in shaping an ecologically healthy and regenerative world in the 21st century," rather to practice "little more than a minor decorative art."[9] Despite the frequent citing of Ian McHarg's assertion that the study of environmental ethics, with its roots in ecology, is absolutely crucial to landscape architecture, very few design degree programs offer an environmental ethics course. A 1992 ASLA survey revealed that only three of forty-three degree programs had ever offered a full-credit course on environmental ethics. This was regarded as not only embarrassing but also outright dangerous.[10] Things have not improved in the intervening decade.

Landscape architecture is often said to advance wise stewardship of the land, yet its degree programs rarely prepare students to do this. James Patchett, chair of the ASLA Professional Interest Group on Water Conservation, has decried the frequent failure of the profession to live up to its ethical responsibility for "the stewardship and conservation of natural, constructed, and human resources."[11] This "failure of contemporary landscape architects to articulate their role satisfactorily as 'stewards of the land'" is due, Robert Scarfo (of Washington State University in Spokane) argues, to a delusion inspired by an antiquated romantic ideal of landscape husbandry completely out of touch with the technology-driven realities of the modern profession.[12]

The debate about the motivations and environmental efficacy of landscape architecture frequently takes place in the pages of the profession's trade journal, *Landscape Architecture,* as do claims about the human and natural benefits of the attention-grabbing projects

The stormwater treatment pond outside the Water Pollution Control Laboratory in Portland, Oregon, represents a beautiful fusion of form and function in sustainable landscape design. Wanting to lead by example, the agency (whose task is to monitor urban runoff) created a demonstration project in which stormwater treatment is made dramatically visible. The rock-lined dissipater that slows runoff inflow and the water-deflecting, storage time–increasing rock wall that prominently rises from the stomacher basin have been featured on the covers of magazines and books about sustainable landscape construction. Nearby nonfunctional art (the token 10 percent-of-costs civic project) pales by comparison. Courtesy of the author.

presented therein. In a recent article about the 2002 ASLA awards, jurors referred to "the dearth of ecologically sensitive designs" from which to pick, the "flawed presence [of ecology] in so much of the work" submitted, and the overall impression that "the profession is only giving lip service" to sustainable design.[13] It appears that little has changed in the decade since Thayer wrote that landscape architecture is "dominated by the creation of pleasant, illusory places which either give token service to environmental stewardship values, or ignore them altogether."[14]

"Architecture," says a prominent critic at the Harvard University Graduate School of Design, is "a destructive act," with the phrase *green architecture* being as oxymoronic as *green* SUVs. The most serious question that can be asked about landscape architecture is whether it too is, overall, environmentally constructive or destructive. How effective the profession is in generating environmental benefits can be gauged by reviewing the projects covered in *Landscape Architecture Magazine*. Luckily, a convenient way to make such an appraisal exists.

One of the most exciting and promising developments that is fostering sustainable design is the increasing use of the U.S. Green Building Council's Leadership in Energy and Environmental Design (LEED) rating system, which evaluates the environmental performance of buildings and sites.[15] A subset of its criteria appropriate to water-sensitive design includes strategies such as minimizing parking spaces, reducing impervious surfaces, installing multiple source stormwater treatment technology such as bioretention swales,[16] building green eco-roofs and rain gardens,[17] and developing on-site water reuse systems.

A systematic review of the past decade of projects covered in the magazine shows a striking absence of water-sensitive design: less than a third explicitly managed water in ways that would give them moderate-to-high LEED water rating credits, and for the remaining two-thirds, the amount of LEED water credits that could be awarded was minimal—less than 10 percent of the potential total.

Based on this sample, "standard" landscape architecture is not "green." Yet many of the projects that earned few or no water-sensitive LEED credits may have offered some marginal water improvements over the previous site conditions, and thus it might be argued that they pose less of a threat to nature than a building—no matter how green its design—would. But given that landscape architects pride themselves in being more environmentally sensitive than architects, it may be that such a self-righteous attitude needs to be tempered. In the end, perhaps the best that can be said is that, on average, the projects published in the profession's primary magazine neither harm nor help nature.

Should such a conclusion surprise us? The most in the know would argue not. The one article in *Landscape Architecture* on the LEED credit system concluded by questioning why landscape architects have been so little involved in developing and applying the system. The answer supports my belief that most landscape architects either ignore the issue of "greenness," or, if they do refer to themselves as "green," are usually in reality gray at heart: "Many landscape architects feel that they design sustainable landscapes as a matter of course in their general practice and that they don't need LEED to guide them. There is also a misguided assumption that all built landscapes are 'green.'"[18]

Of course, as even my admittedly small sample showed, such arrogance is unwarranted and instead supports Thayer's contention that "most products of landscape architecture are simply *not* sustainable by any definition."[19] In Thompson and Sorvig's review of over a hundred sustainable landscape projects (selected based on their profession of "sustainability"), they grapple with the troubling reality that these landscapes sometimes *harm* the environment.[20] Never, they note, should we forget that no matter how naturalistic or sustainable a created landscape appears or is touted to be, it is not a substitute for nature free from human meddling. An exception might be made, however, for landscape restoration projects such as stormwater wetlands designed to improve water quality, reduce floods, and enhance wildlife habitat.

"Functional Art," the Key to Success in Sustainable Site Design?

Given that, in Thayer's words, "the majority of the work done by [landscape architects] . . . could not possibly be justified under official ASLA rhetoric pertaining to environmental stewardship or sustainability," and that perhaps the best that we can ask from any site design project is that it "tends" toward sustainability, are there projects that transcend the norm?[21]

The single most effective action that can be accomplished for the future of nature is to motivate and inspire large numbers of people. If enough people cared enough, needed reforms would be put in place. (Carl Steinitz argues that only fear is an effective motivator.[22] But there have been plenty of proposed alterations to environments halted because people loved what existed.) Motivation will come from people's experiences of relatively undisturbed, protected green spaces far from cities, but also from educating and directly engaging people in the recognition and repair of damaged landscapes. Whereas the former is the purview of conservation biologists and nature writers, the latter is very much the business of restoration ecologists and landscape architects. Through melding engineering and aesthetics, developing what might be called "functional art," landscape architects can contribute to sustaining nature. The reason for this is that neither art and design nor science and engineering alone have done much to instill love of and motivate action for the natural world. No one would be inspired by a sterile, engineered waterway (like the Los Angeles River) to protect other rivers, just as no one would become dedicated to preserving rain forests because they contemplated a tree clipped to look like a giant puppy.

The quotations posted on my office door have garnered coverage by *Harvard Design Magazine*. One is from Christopher Caudwell and was cribbed from landscape architect Garrett Eckbo's once influential *Landscape for Living*. There may be no better challenge anywhere to C. P. Snow's assertion that art and science inhabit different worlds: "Art is the science of feeling. Science is the art of knowing. We must know to be able to do. But we must feel to know what to do."[23] The pressing question becomes can the *feeling* of art and the *knowing* of science be married through landscape architecture as a means for sustaining nature? The answer is a qualified yes, as shown perhaps most clearly in the recent development of functional *and* beautiful stormwater wetland parks.

Wetlands combine beauty and ecological function in a way that few other landforms can. As such, they have been and will continue to be important elements in site design and landscape planning. There is a long tradition of scenic wetland gardens. Indeed, landscape design probably began with the publication of Toshitsuna Fujiwara's eleventh-century *Sakuteiki,* with its instructions about how to build Japanese water features.[24] And modern landscape architecture is often thought to have started with Frederick Olmsted's work on Boston's Back Bay Fens wastewater treatment park system. Since then, wetlands have been constructed primarily by engineers and scientists for flood prevention and water quality improvement. Though these wetlands have functioned well, their generally square shapes have provided little benefit to wildlife and have been aesthetic ciphers. But the synthesis of art and science has nowhere been more successfully accomplished than in the creation by landscape architects of treatment wetland parks that, in acknowledgment of Olmsted's previously neglected vision, combine environmental management and ecotourism.

The trend away from single-purpose treatment wetlands and toward multifunction designed wetland *parks* is the success story in nature-sustaining landscape architecture. No longer are ecological features like wildlife habitats or human amenities like education centers treated as ancillary; instead, they are acknowledged to be as important as water management. The projects illustrated here and in my *Wetland Design*, arranged in order from naturalness to artifice, have won numerous awards and are worth briefly introducing as examples of visionary built wetlands, strong in both function and form.[25]

All these projects improve the ecology of their immediate surroundings. And since both insults to and purifications of water are additive and transferable to the larger landscape, these site effects are felt downstream and help sustain the entire watershed. In their beauty, these created wetlands also inspire activism for the protection of natural wetlands elsewhere.

Although what in sustainable ecological design constitutes the "right" balance between nature and artifice (function and form) is debated, these projects show that one need not dominate at the expense of the other and that the extent of the designer's imprint on the land can successfully vary. There is no real conflict between form and function. And, as Thompson and Sorvig note, we usually find nature's own functional forms to be supremely beautiful.

Functional art lies at the success of ecologically sustainable designs that will inspire action beyond the bounds of the site. Louise Mozingo, associate professor in the Department of Landscape Architecture at the University of California, Berkeley, is right to argue that no matter how righteous ecological design projects make one feel, their frequent aesthetic insensitivity sends viewers fleeing to the nearest Italian garden.[26] It need not be this way. The moving poetry and haunting beauty of gardens like those of Kyoto or Suchzou can be inseparable from the engineering of modern water treatment and stormwater management. Thayer is on target again: "Sustainable landscapes need conspicuous expression and visible interpretation, and that is where the creative and artistic skills of the landscape architect are most critically needed."

Continuing, Thayer succinctly concludes, "But the new institutions needed for a transition to a sustainable world must ultimately be based upon the perception and comprehension of the ordinary people who will create them. In turn, *their* ultimate reality is in the land and spaces around them. The small steps taken to build sustainability into the local landscape in discreet, manageable chunks which people can observe, try out, experience, and improve are actually large steps for humankind."[27] Amen.

2003

Notes

1. http://www.worldtrans.org/whole/warning.html.
2. At the first National Green Building Conference in Austin in Novem-

ber 2002, organizers had to turn away interested potential attendees after the first several thousand were admitted; obviously there is an incredible hunger out there for learning about these matters.

3. Robert L. Thayer Jr., "The Experience of Sustainable Design," *Landscape Journal 8*, 1989, 101.

4. John T. Lyle, *Design for Human Ecosystems: Landscape, Land Use, and Natural Resources* (Washington, D.C.: Island Press, 1985); Robert Thayer, *Gray World, Green Heart: Technology, Nature, and Sustainable Landscape* (New York: John Wiley & Sons, 1994).

5. J. William Thompson and Kim Sorvig, drawings by Craig D. Farnsworth, *Sustainable Landscape Construction: A Guide to Green Building Outdoors* (Washington, D.C.: Island Press, 2000).

6. Craig Campbell and Michael Ogden, *Constructed Wetlands in the Sustainable Landscape* (New York: John Wiley & Sons, 1999).

7. http://www.tcaup.umich.edu/facultystaff/sustdesign.html.

8. From an unpublished Loeb Fellowship document.

9. ASLA, "Taking Up the Challenge," *Land 35*, 1993, 5.

10. ASLA, "Environmental Ethics: Elective Only?" *Land 35*, 1993, 2.

11. James Patchett, "Letter from the Chair," *ASLA Professional Interest Group on Water Conservation Newsletter*, Spring 1999, 6.

12. Robert Scarfo, "Stewardship and the Profession of Landscape Architecture," *Landscape Journal 8*, 1989, 60–68.

13. "ASLA Awards," *Landscape Architecture*, November 2002, 66.

14. Thayer, "The Experience of Sustainable Design," 103.

15. *LEED Green Building Rating System*, Version 2.1, http://www.usgbc.org/Docs/LEEDdocs/LEED_RS_v2-1.pdf.

16. Bioinfiltration swales are berms, small dams, or depressions created by excavation placed in channels intended to filter the first half-inch of stormwater runoff from impervious surfaces through a grass or vegetative root zone.

17. Rain gardens absorb stormwater from roofs, thus reducing its flow off-site.

18. Meg Calkins, "Leeding the Way: A Look at the Way Landscape Architects Are Using the LEED Green Building Rating System," *Landscape Architecture*, May 2001, 36–44.

19. Thayer, "The Experience of Sustainable Design," 102.

20. Landscape architecture projects can cause environmental degradation in several ways, including preventing water absorption (and thus lowering water tables and allowing polluted runoff into storm drains, rivers, and streams), reducing biodiversity, introducing non-native rogue species, establishing landscapes that use ecology-disrupting fertilizers and pesticides, and so on.

21. Thayer, "The Experience of Sustainable Design," 102.

22. Carl Steinitz et al., "What Can We Do," *Harvard Design Magazine* 18 (Spring/Summer 2003).

23. From Christopher Caudwell, *Illusion and Reality* (New York: International Publishers, 1947), as referred to in Garrett Eckbo, *Landscape for Living* (New York: Architectural Record, 1950.)

24. Jiro Takei and Marc P. Keane, *Sakuteiki, Visions of the Japanese Garden: A Modern Translation of Japan's Gardening Classic* (Boston: Tuttle Publishers, 2001).

25. Robert France, *Wetland Design: Principles and Practices for Landscape Architects and Land-Use Planners* (New York: W. W. Norton & Company, 2002).

26. Louise A. Mozingo, "The Aesthetics of Ecological Design: Seeing Science as Culture," *Landscape Journal*, Spring 1997, 46–59.

27. Robert L. Thayer Jr., "Gray World, Green Heart" in *Theory in Landscape Architecture: A Reader,* ed. Simon Swaffield (Philadelphia: University of Pennsylvania Press, 2002), 189.

11

Green Good, Better, and Best: Effective Ecological Design in Cities

Kristina Hill

Landscape architects and other ecological designers are creating new prototypes of landscapes that contribute to ecological health in cities worldwide and are often doing it quite well. But the professional magazines that showcase design excellence, *Landscape Architecture, Land Forum,* and others, rarely discuss or clarify what makes some ecological designs better. Typically, reviews in these aesthetically oriented publications stress the way places and things look over how they function biologically, and only infrequently ask whether the ecological strategies used in the designs are contextually appropriate. Even more important, they do not ask if these designs have the potential to be implemented widely enough to make a broad difference to the state of urban ecosystems.

In this short essay I cannot hope to make up that deficit, but it seems reasonable to hope that I can offer a starting point. I will use examples from two cities where I am aware that effective ecological design has been implemented to build my argument that if we reflect carefully on our choice of criteria, we can distinguish "good" ecological design from that which is merely wishful or stylishly iconic. So that the reader can keep score, I will present my criteria up front.

First, "good" ecological design must enable a biological function important to the ecological health of its regional and local setting.

UFA Fabrik, Berlin: outdoor café with roofs visible. Photograph by Martin Bjork.

"Better" ecological design would address the functions that are strategically critical to this health in a given bioregion—such as improving water quality, conserving plant or animal species at risk of local or global extinction, increasing soil fertility, or improving air quality. "Best" ecological design would be able to show these benefits in measurable ways and would stay in touch with the latest thinking in the sciences and engineering (and social and environmental ethics) to make sure it makes sense to keep pursuing those benefits.

Second, "good" ecological design has to be more cost effective than our existing methods of addressing (or ignoring) urban ecological health. Designers are not going to be able to promote new prototypes if they are all a collection of expensive "demonstration projects" whose most lasting lesson is that we cannot afford to change. Cost-effective solutions can—and likely will—be implemented widely. Individual site-scale designs will not have much impact on the regional and even global ecological trends that matter most. Replicability ties ecological design to the concept of infrastructure as used by planners, economists, engineers, and even the new theorists of landscape urbanism. If a prototype is cost effective, socially useful, politically palatable, and aesthetically pleasing, it can be propagated through urban space—just like the American lawn or the ubiquitous stormwater pipe and catch basin system.

Third, "good" ecological design has to be elegantly parsimonious. By that I mean the same physical forms (structures, plantings, and topography) must fulfill both ecological and social needs. Forms have to be associated with cultural meaning that is engaging and valued. Some would say that is just the essence of good design. Maybe so. But if a design proposal does not meet the first and second criteria, I am arguing that it is not good *ecological* design. I am betting most designers will not want to be held to those two criteria, so it seems safe to say we will still need to talk about other kinds of "good design" that (a) do not address critical ecological functions and (b) are not more cost effective than our current norms. While it certainly is important for us to learn from a wide variety of experimental projects, the three criteria I have presented are useful for keeping score as I discuss examples from two cities that figure prominently in urban ecological design: Berlin and Seattle.

Berlin

The capital of the new Germany is a city that for decades has been regarded in Europe as a hotbed for experiments in architecture, urban design, and the politics of new world orders. It is not surprising therefore that it is a place where a large number of experiments in housing, park, and civic space design have claimed to be ecologically strong over the past three decades.

In August 2002 I lived with a dozen students at the former Universum Film company (UFA) factory site in Berlin, which is famous for producing the 1927 futurist film *Metropolis*.[1] Today it is an active community center where Berlin residents and visitors are seduced into learning about sustainability by an active performing arts program that brings in visitors, who see green roofs and solar panels while enjoying cabaret and outdoor dining. Starting in the mid-1980s, experimental green roofs have been built on many of the original buildings. For almost a decade, landscape architect Marco Schmidt and his research associate, geographer Katharina Teschner, have been monitoring the roofs' vegetation, microclimate, water-quality impacts, and stormwater detention. They find that the roofs function pretty much as planned. The plants (and thin soils) hold 95 percent of the rainwater that falls during the season when stormwater could otherwise flood sanitary sewer lines, cool the air above them by evaporating water through their leaves, and have been increasing in diversity through natural seed dispersal.[2]

Schmidt and Teschner also monitor the green roofs at a very different site in Berlin—the glitzy urban futurists' dream at Potsdamer Platz. They helped design the plantings for the DaimlerChrysler headquarters buildings, master planned by Renzo Piano and including a landscaped plaza designed by Herbert Dreiseitl. Sedum and allium species were the rooftop plants of choice, since they survive the harsh exposure conditions and grow well in thin soil. After only a few years, the roofs have filled with plants, and sedum species coat them in rusty shades of orange and green. Simple, green, effective, and affordable, Germany's green roofs offer a performance-tested model of ecological design ready for adoption in North America. A financial benefit is that they last longer than "normal" roofs, since they are protected from extremes of heat and cold. Ecologically, the plants store rainwater and increase the amount of water that evaporates to humidify local air. Berlin ecologists have identified the decrease in evaporation by plants as an impact of urbanization that significantly damages the remaining urban vegetation and local microclimates. The "sponge effect" of green roofs is also a critical function in Berlin as in many contemporary cities where stormwaters flood sanitary sewer pipes and force raw sewage into urban waterways. Green roofs meet the financial interests of long-term property owners thanks to reduced stormwater sewer rates and density bonuses in Berlin's building codes.

At ground level on the DaimlerChrysler site, landscape architect Herbert Dreiseitl's water pools present a different story. Apparently, the clients wanted a reflecting pool to showcase their buildings. Dreiseitl argued that the pool should store stormwater to save the cost of building storage vaults underground and still take advantage of Berlin's financial incentives for stormwater detention. The pools are beautiful and engaging for people walking through the site during all four seasons. For some designers, using water in a way that engages people is enough to qualify as successful ecological design. But the ecological story under the aesthetic surface offers lessons on urban ecological design that are strategically critical. First, the good news. The ponds were originally filled with drinking water to establish the desired water quality. Whenever there is too much rain, the pools simply overflow into one of Berlin's canals. Supplied only by recirculated rainwater, the pools lure people to walk around them, have lunch sitting on their rock edges, and dangle in their feet. People even put ornamental goldfish in the pools. Clumps of aquatic grasses create the aesthetic impression of an urban wetland.

Now the bad news. The absorbent green roofs and the rainwater-using toilets inside the buildings leave little roof water for Dreiseitl's pools. The rainwater available comes as runoff from the paved walkways around the pools' edges, plus a bit of occasional roof overflow. This walkway runoff washes a fair amount of nutrients into the pools, and birds contribute more. These nutrients cause algae to grow rapidly, so the water must be filtered. The existing clumps of wetland plants cannot naturally filter all of these nutrients. In this regard, Dreiseitl's pools do not function like wetlands even though they look like them.

To get rid of the nutrients and algae, a complex mechanical system lies beneath the DaimlerChrysler pools; it pumps the water into a cistern, then filters and recirculates it into the pools. DaimlerChrysler pays the significant costs of this mechanical filtration. Water flows are tracked in real time by computers that make the underground room look like a busy little wastewater treatment facility. Resembling the workers' city that lies below Fritz Lang's cinematic *Metropolis*, full of sweating men and steaming machines, this ecological structure depends on a hidden use of energy and capital that must be fed—in this case, with money.

The pools and their underground filtration system offer an impressive show of money and muscle and can be seen as evidence of Daimler-Chrysler's commitment to environmentally responsible design—at least until you find out that in 2008 ownership of the pools and their underground machine rooms—along with the responsibility to maintain them—will be transferred to the City of Berlin. And Berlin, for those who have not been keeping up with municipal finance trends, happens to be broke after a decade of reunification parties and efforts to attract investors. Odds are good that this stormwater filtration system is not going to be a high priority in a city with no budget to maintain existing parks and museums.

My point is that expensive projects like this will probably not contribute to urban sustainability. The costs of mechanical filtration and recirculation and the maintenance of the machines that do the job are much higher than the costs of ordinary stormwater drainage. If there were a cost-effective way to use high-volume mechanical filtration in highly developed urban sites, I would draw a different conclusion.

But since there is not, no matter how good this project looks, it is not a design that should be replicated.

In general, the best examples of ecological urban design that we saw in the rest of Berlin involved very sophisticated energy-saving materials and designs, effective urban transit systems that provide viable alternatives to driving, and green roofs. Berlin's relatively low energy use and high air quality are exemplary. The natural lighting and ventilation in Berlin's buildings are also exemplary, but they rely on Germans' tolerance for a wider range of indoor air temperatures than Americans currently accept.[3] It was disconcerting to learn that the general public is not as involved (because of limited governmental outreach) in this type of planning and design as they usually are in similar American projects. This disengagement may undermine the Berlin government's efforts in coming decades. In addition, many ecologically minded Berliners seem interested exclusively in the biodiversity of plants. While Europeans might argue that there is too little animal diversity left to give a hoot, the absence of attention to the full range of biodiversity means that Berlin's model of ecological design is not appropriate for international export.

Seattle

Halfway around the planet at about the same latitude is a city with a politically similar climate: Seattle. People there have long supported civil protest and local decision making and have wanted their city to wear a deeper-than-average shade of green. And while its public coffers are not as empty as Berlin's, it has the same economic woes as other American cities with jobs in NASDAQ companies or airplane factories. Ecological design there, as in Berlin and just about everywhere else, has to make financial sense to be implemented in more than demonstration projects.

Seattle has an unusual biological context for its urban growth, perhaps partly because its region was urbanized only in the last half of the nineteenth century. Three-foot-long salmon still swim into urban waterways from the ocean to find smaller streams where they can spawn, then swim out again as small fry. Harbor seals glide along its urban shoreline, bald eagles nest in public parks, and cougars live a few miles outside the city limits. But Seattle is sprawling, and its residential density—about six thousand persons per square mile—

remains lower overall than that of Los Angeles. Not coincidentally, several populations of salmon were listed as threatened under the U.S. Endangered Species Act a few years ago, making this city and its region (along with Portland, Oregon) the first urban test cases for that federal law.

Much of the most interesting work in ecological design in the Seattle area has been done by or for the public sector. Since the early 1990s, the city and county have worked to manage stormwater in an increasingly decentralized system of detention, filtration, and infiltration devices. In the city's case, these devices are often designed to also function as civic spaces. One of the early experiments, Meadowbrook Pond, involved building a public park around a shallow artificial pond to collect flood overflows from an urban creek. There is no shortage of such overflows, since an increase in the area of upland parking lots provides far more rainwater runoff than Seattle's pre-development landscape would have funneled to these creeks. While this park succeeded in bringing public art and infrastructure together, it was less successful at establishing nonhuman habitat. The level of water in the pond is too deep to allow much aquatic vegetation to survive, so it has become a rather expensive duck pond not particularly suitable for conservation of species that are in trouble, like salmon and trout.

Meadowbrook Pond functions pretty well as a neighborhood park. It is a valued and used public space. But the millions of dollars it cost went toward only a few acre-feet of stormwater storage. So, like Dreiseitl's mechanical pools at Potsdamer Platz, it is not a prototype that can be replicated easily enough in other parts of the city's infrastructure system to make a real difference in ecosystems.

The Emergent Ecology of Urban Streets

Since the Meadowbrook experiment, Seattle Public Utilities (SPU, the city's solid waste, water supply, and drainage department) has adopted a different strategy for stormwater-related projects. It has continued to experiment with ways to decentralize stormwater detention, but increasingly these designs are located in the street right-of-way zone, not in park landscapes. SPU's civil engineers and landscape architects, as well as strategic policy staff of all professional stripes, have been developing street designs that are affordable as well as ecologically and socially beneficial. Their local biodiversity goals are modest, consisting mainly of sustaining native plants. But their goals for the cumulative impacts of these projects are truly ambitious and can directly impact the potential for salmon to survive in Seattle and its region. In effect, they are trying to transform how streets function ecologically. And since public right-of-way land makes up about 38 percent of Seattle by area, these new prototypes can make a significant difference in the larger urban fabric.

Two particular street experiments, known as Viewlands Cascade and SEAStreet (where SEA stands for Street Edge Alternatives), have already been built and monitored for more than two years. They were constructed in a part of Seattle where rainwater historically drained to open ditches lined with grass or concrete and from there flowed into creeks or larger water bodies. These projects have social, hydrological, and fiscal implications for ecological urban design. First, people say they like the new informal native plantings and the way that horizontal curves in these streets slow traffic. These curves have allowed designers to fit in more stormwater runoff detention in swales and bowl-shaped depressions in the lobes of the curves. Residents also say they are willing to assist in maintaining these native plants even if they have to use their own money.[4] Second, the street designs turn out to be more than capable of detaining the volume of a significantly worse-than-typical rainfall during Seattle's wet winters; this was the designers' goal. Water-quality benefits are already evident and seem to result from the mechanical filtration accomplished by plant shoots, as well as (in the case of Viewlands Cascade) from infiltration into soils and subsurface glacial deposits. The development of one block of a "natural drainage system" like SEAStreet costs only two-thirds of what it would cost to make standard street drainage improvements (adding underground pipes, catch basins, and sewer inlets) in these ditch-and-culvert neighborhoods. What Seattle is doing is quietly radical, fiscally strategic, politically positive (residents of these neighborhoods are getting their first sidewalks!), and incremental. Most important, these street designs seem likely to provide the improvements in both water quality and stream flow levels needed to keep salmon swimming in urban creeks.

Urban design and ecological design would rapidly part ways, however, if ecological performance could be improved only in low-density residential neighborhoods. Seattle's approach to urban street ecosystems also includes designs for late-nineteenth-century-looking streets for higher-density settings where there is no room for curving berms and swales. The highest density example is a 120-acre Hope VI public housing redevelopment, designed with an average of sixteen units per acre. This project, known as High Point, is in its final stages of permitting. Fifty percent of its units should be built and occupied by 2005.[5] The design of better-performing streets for higher-density urban neighborhoods provides a critical piece of evidence for the claim that the inclusion of ecological design strategies does not have to result in less affordable (or less dense) urban residential or mixed-use neighborhoods. Some fascinating lessons can be learned from this case about the interactions of natural processes, city politics, and project design teams—all the nitty-gritty that is fit to print, in fact. But I will focus now on the nits and grits that help clarify what is needed for effective ecological design.

The High Point project has its origins in urban design rather than engineering. It began as an exploratory graduate studio in the University of Washington's Department of Landscape Architecture and will continue as a monitoring project in collaboration with engineering faculty. When SPU staff saw the students' work, they were inspired to pursue it as one of their flagship investment projects. The SPU project manager for High Point, Miranda Maupin, was trained as a landscape architect, and the civil engineering firm doing most of the design work (SvR Design) is a partnership between a landscape architect and a civil engineer. The fact that broad design thinking was its starting point has implications for what the design professions can do to improve their approach to ecological design as infrastructure and urban fabric, not just isolated site experiments.

The key to ecological design at High Point was the insight that the social goals of the federal Hope VI low-income housing program can be compatible with the ecological goals of a housing area that drains into an important salmon stream (Longfellow Creek, which gets more adult salmon returning from the ocean than any other Seattle stream). The area of the redevelopment adjacent to this stream constitutes 10 percent of its drainage basin. That geography made this project a strategic ecological investment for Seattle.

Hope VI mandates that its public housing projects must resemble their surrounding residences so that those living in public housing are not physically or perceptually isolated from their neighbors. In other words, these redevelopments cannot be a "ghetto" of any kind, not even a green one. The surrounding neighborhoods date from the early twentieth century. Streets are a narrow twenty-five feet of paving that include two-way travel and parking on both sides. The public right-of-way includes an eleven-foot-wide planting strip and five- or six-foot-wide sidewalks on both sides. Using this prototype, the project team adapted the student design ideas to a rigid set of dimensions that allowed for no wasted space. The narrow paved width of the street was at first opposed by fire department officials, who argued for the contemporary subdivision standard of thirty-two feet. Here the Federal Hope VI program's contextual design requirement was politically useful: it allowed the project team to argue for narrow streets for both social and hydrological reasons. The next big question was where and how to detain and filter the stormwater off those streets and the roofs of the planned housing units. The planting strip began to look awfully interesting.

The final design proposal was to use this strip to store and filter street and sidewalk runoff water in ways largely invisible to casual observers. This strategy allowed focus on detaining and filtering the typical small storms, which recent science has identified as critical to fish health downstream, where the runoff waters end up. It also focused on managing stormwater flows and pollutants at the block scale and on allowing the block-level designs to add up to a network of storage and filtration at the neighborhood and sub-basin scales.

To establish a basis for comparisons, the High Point project team took the unusual approach of maximizing detention capacity and then modeling detention performance in a hypothetical predevelopment landscape. Their customized model uses continuous rainfall simulation to compare the predicted hydrologic performance of the final housing project design to the hydrologic performance of 120 acres of undeveloped pastureland. This simulation work represents an important step in ecological design—the use of "best available engineering" to turn the hypotheses of good designers into good experiments as site construction is completed and monitoring results start to come in. To improve our ability to address ecological functions and not just make cities look greener, we need to figure out how to simulate the performance of these designs and then test those simulations using monitoring data. Without clear performance expectations, explicitly modeled conditions, and actual performance data, it is going to be hard to work within the narrow range of risk most clients can accept and still learn to do this kind of design work better.

The High Point project team had to do some innovative fiscal planning to build critical interdepartmental alliances. SPU will put up to $2.5 million of its own capital budget on the line to build this natural drainage system, even though it is a project that technically belongs to a different entity, the Seattle Housing Authority. To secure the cooperation of the Housing Authority, SPU guaranteed that the new system would work or they would pay to replace it. One benefit of this arrangement is that the innovative technology of the "planting strip as stormwater facility" is going to be owned by the same entity that designed and will maintain it. Lessons learned will accrue to the benefit of the owner, who also owns and maintains the rest of the city's drainage system. That may seem simple, but it is a crucial link in the chain that might allow innovations to spread through the city.

In comparison to cities like Berlin, Seattle is doing "good" ecological design for water quality and aquatic biodiversity. But in the bigger "urban ecological picture," it is only now developing a citywide transportation alternative to the private car, and its air quality is still declining.[6] Moreover, although Seattle leads most American cities in recycling solid wastes, it still falls far behind Berlin in the use of sustainable and efficient building energy systems such as photovoltaics. And ironically, while Seattlites rely on relatively clean hydropower for their electricity, the same dams that generate that power often block salmon from reaching spawning grounds. Even so, at least within the city itself, ecological design work is good and getting better, thanks largely to the leadership of the public sector.

The recent focus on the ecology of infrastructure systems in Berlin, the cities of the Pacific Northwest, and elsewhere supports the very real possibility of eventually creating a new urban ecosystem. In my view, that is the central challenge that ecological design must accept in all cities, if it is going to achieve anything of real importance.

2003

Notes

1. To learn more about this community facility and its programs on sustainability, see http://www.ufafabrik.de.

2. Marco Schmidt and Katharina Teschner can be reached care of the UFA Fabrik, Berlin, Germany, using the fax number 49 30 755 03 110.

3. I am indebted to Lynne Barker of the U.S. Green Building Council, one of the developers of the LEED project rating system, for pointing out to me the importance of this greater tolerance.

4. Data on public perception of the SEAStreet design were collected and analyzed by Melanie Mills in her master's thesis, "Alternative Stormwater Design within the Public Right-of-Way: A Residential Preferences Study," in landscape architecture at the University of Washington, Seattle, 2002.

5. The SPU project team for High Point is led by Miranda Maupin, senior planner (miranda.maupin@seattle.gov), and includes Tracy Tackett, senior civil engineer, and Ray Hoffman, director of strategic policy. The design team at SvR Design that is working with SPU is led by Peg Staeheli, ASLA. The hydrologic model is being developed by Robin Kirschbaum at the engineering firm of R. W. Beck, Seattle.

6. Seattle voters approved a fourteen-mile regional light-rail line in 1996 and a fourteen-mile in-city monorail system in November 2002, at a combined transit infrastructure investment of about $4 billion.

12

Energy, Body, Building: Rethinking Sustainable Design Solutions

Michelle Addington

During the past two decades, initiatives devoted to understanding and controlling the human contribution to environmental degradation have shifted from the provenance of environmental activists to state institutions and international coalitions. Treaties, codes, policies, and agreements have resulted, and few activities in the public realm can take place without at least a nod to environmental issues. Research centers and university programs have emerged to comprehensively address such aspects as atmospheric and geophysical monitoring, and grassroots efforts in communities have produced recycling programs and heightened public awareness. No other problem during the twentieth century mobilized the public and private sectors across such a wide swath in such a short time. Ready solutions, however, have not been so forthcoming.

Environmental degradation is among the most difficult and complex problems ever faced by modern society. Given the current state of discourse in architecture, one would not think that this is the case. Notwithstanding the fact that discourse about the environment is omnipresent, the concerns of architecture are clearly of a different ilk. Most architects believe that the solutions are known and straightforward and that their implementation requires only an act of will and the commitment of a few extra dollars. Within both academia and professional practice, the focus has been placed on communication: instructing students in these solutions or "best practices" and convincing clients and architects to share the ethical stewardship necessary for implementation.

The terms *green building* and *sustainable design* were originally coined to name the collection of strategies that we generally accept as environmentally responsible, but their use has been rhetorically expanded to include intentions as well as actions. Firms, consultants, and manufacturers use the terms to distinguish themselves from their competitors, and few clients would be willing to consider funding projects that are lacking in aspects labeled "green." Recently, "Teaching Green," an article in a design magazine, suggested that architecture schools could even be ranked by how many courses "employ the words *environment, green, sustainable* as key descriptors."[1] Indeed, in comparison with many other disciplines that also have environmental consequences, architecture may well rise above the crowd in its many activities demonstrating a profound concern for stewardship. Noble intentions, however, do not always lead to effective results.

Intentions and actions aside, energy use and greenhouse gas emissions are continuing to rise to unprecedented levels. In 2000, the United States consumed 45 percent more energy than it did in 1970 and is expected to consume 93 percent more by 2020.[2] This measured and projected data correspond to time periods in which numerous initiatives for conserving energy had been or were projected to be put in place. Furthermore, emissions from the building sector are increasing at a rate faster than those from the other sectors (transportation, industry, and agriculture). Currently, buildings are responsible for over one-third of the nation's total energy use, and two-thirds of its electricity use. Some of the most disturbing data were released from the Department of Energy (DOE) in a 1999 study of commercial buildings, documenting that newer buildings, even though they generally reported having more energy-efficient features, used more electricity than older buildings.

The problem is not so much that current energy conservation initiatives are flawed, but that they do not consider the most significant determinant of building energy use—size. Newer buildings tend to be larger than existing buildings, with more square footage per occupant and per function. Further exacerbating energy use, additional space in a building increases the energy use of the ambient systems— lighting and HVAC—by as much as the square of the added floor space. For example, other data from the DOE in 2000 project that although the number of households is expected to increase by 1 percent a year, the residential energy demand will increase by 1.9 percent, while an increase in commercial floor space of 1.3 percent will produce a 2 percent increase in electricity use. Energy reductions wrought by efficiency improvements are quickly subsumed and surpassed by the energy demands to support the additional space. The self-evident conclusion seems to be a rather dismal one for architects: we must reduce the size of new construction. Notwithstanding the difficulty of implementing and enforcing such a measure, it would have no impact on the existing building stock. Furthermore, the increasing size of buildings is a trend not just in the United States; in China, residential space per capita has more than doubled since the 1960s and is expected to quintuple during the next decade.[3] We can only expect energy use and greenhouse emissions to continue their precipitous climb.

For all of the guidelines, standards, and assessment systems rapidly being adopted by and assimilated into architectural practice, minimal attention has been devoted to understanding the roots of building energy use. We have bracketed the problem as occurring within our normative practice, focusing on improving the efficiency of the systems and materials that we routinely use. We ask which insulation has the highest R-value to reduce the heating load, and we specify luminaires with the highest efficiency fluorescent lamps. In working to select the best equipment and strategies to improve the performance of our standard systems, we never stop to ask how heat and light behave, and we certainly do not ask the most fundamental question: for what purpose do we heat, cool, or light a building?

Most of the major energy-consuming systems in a building, including lighting, heating, cooling, and ventilation, are intended to service the needs of the human body. Our conventional systems, however, are designed to service the building. We heat and cool its entire volume; we provide standard lighting levels throughout its rooms. As buildings become larger, smaller and smaller percentages of these systems are devoted to meeting the needs of the human body. Yet, the body's heat exchanges occur within a zone of a few centimeters around it, and the eye intercepts only a tiny fraction of the light in a room. Our conventional systems provide ambient conditions in a building—a steady seventy-two degrees Fahrenheit or a constant forty foot-candles. The body, however, is insensitive to

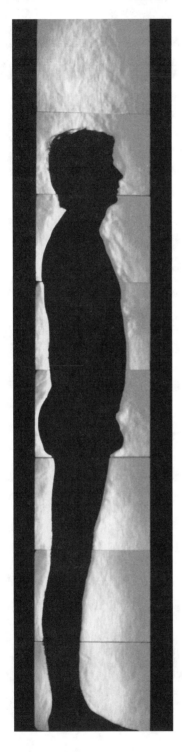

homogeneous conditions; its neurological system responds only to difference and change in the environment. If the body feels cold, it is not due to the surroundings being cold. The body is a heat-generating entity, and as such it is always exchanging heat with its surroundings. To maintain homeostasis, that exchange should be "steady state"—the surroundings must receive the same amount of excess heat that the body is producing—any more and the body is losing heat, and any less and the body begins to overheat. Rather than sensing the temperature of the surroundings, the body is sensing the rate at which its own heat exchange is changing—if the rate is increasing, the sensation reads as cold, if it is decreasing, the sensation reads as hot.

The eye has several complex mechanisms for seeing, all based on relative differences of stimuli within the field of view. The difference may be in luminance (light levels), color, texture, orientation, or movement. The eye does not recognize any of these stimuli in absolute terms, only in relationship to other stimuli in the field of view. Like the rest of the neurological system, the eye is insensitive to constancy and cannot differentiate between a steady high light level and a steady low light level.[4] We can

Schlieren photograph of naked man in profile, showing the column of rising hot, turbulent air. Copyright Dr. Ray Clark/Science Photo Library.

only determine light levels comparatively. A surface with a low light level will appear dark if placed next to a surface with a higher light level, but light if placed adjacent to a surface with a lower light level. Although lighting standards are written in absolute terms, one of the few absolute levels truly necessary for vision is the threshold level—the minimum necessary light level that must be present for visual performance. This is quite low, approximately 0.3 foot-candles.[5]

Once the light level is above that threshold, the primary mechanism for seeing is relative difference. Differences in the field of view can be quantified as contrast. A contrast of three to one between the lightest surface and the darkest surface in the field of view is considered quite good for distinguishing objects and details. When contrast ratios are above twenty-five to one, then the eye is typically in a glare situation, and when they are less than about two to one, then the visual performance drops off significantly. Most of the guidelines for lighting design call for absolute light levels that are one hundred to ten thousand times higher than the threshold. For the eye to be able to distinguish objects at such high light levels, then, contrasts within the field of view must be reduced. As a result, the higher the light level, the more even the light must be, which in turn requires the provision of more light in the room to eliminate any dark spots that would produce unacceptably high contrast. Essentially, the focus on an absolute rather than the relative level of light has resulted in an unending spiral of increasing lighting demands.

Many of the guidelines determining the ideal light levels were first codified in the 1950s, when ambient illumination had become the norm after the spread of fluorescent lighting. When fluorescents were first introduced in 1939, they provided little advantage over vapor lamps for factory use or over incandescents for office and residential use. Nevertheless, by 1942 about thirty-three million fluorescent lamps had been sold, and they have since become the standard lighting element throughout the world. The rapid spread of fluorescents may well be attributable to one of the savviest marketing campaigns in the twentieth century, and unfortunately, one in which architects were complicit. Without a compelling reason to switch to fluorescents, customers had to be convinced that the new lamps filled an unmet important need.

Sales manuals from the early fluorescent manufacturers were heavily responsible for encouraging the adoption of both ambient lighting and higher light levels, and it was the higher light levels that necessitated the replacement of incandescents with fluorescents, while the push for ambient lighting supplanted vapor lamps.[6] Factory owners and industrial users were warned that uneven light or "roller coaster" light reduced productivity while increasing accidents, and were encouraged to install ambient lighting systems to blanket the entire building with a constant high level of light. Offices and residential customers were a harder sale, since the adoption of fluorescents required additional electrical infrastructure and thus substantial investment. In perhaps a more duplicitous strategy, owners and architects were encouraged to lighten the colors of the surroundings to accommodate the increased light levels, which tended to increase contrast ratios to unpleasant levels. With a switch to lighter-colored furniture and whiter wall surfaces, the overall lighting level would have to be increased to reduce the contrast ratio. Customers were first convinced that they needed more light, and then the interiors were modified to require that light level.

Today, electric lighting is responsible for as much as half of a commercial building's electricity use and is estimated to be responsible for 20 to 25 percent of the nation's electricity use. One of the widest-spread energy conservation strategies involves replacement of incandescent lamps with compact fluorescent lamps. With efficiencies approximately four times greater than incandescents, fluorescents often are central to the frontline strategy in new energy-efficiency initiatives. This would seem to make sense: relamping requires little infrastructural work, and capital costs are significantly lower than for major equipment overhauls. Relamping is a relatively painless and seemingly effective way to "do the right thing."

The production of light from electricity, however, is an "uphill" energy conversion and thus a process in which the maximum achievable efficiency, constrained by thermodynamic law, is quite low. If one examines total energy conversion efficiency—from coal mine to lamp—then both fluorescents and incandescents operate at net efficiencies below 7 percent. Almost all of the discussion regarding lighting efficacy has used stage efficiencies (in this case, the energy conversion within the luminaire) rather than cumulative or total efficiencies. As a result, eliminating one incandescent lamp has approximately the same impact on reducing greenhouse gas emissions as replacing twenty incandescent lamps with fluorescent lamps. The majority of efforts to curtail the energy use by lighting have focused on efficiency rather than consumption, even though improvements toward the theoretical limit are producing marginally smaller returns.

If we could begin to think about lighting at the small scale—what the eye sees—and not the large scale—the building space—we could drop lighting levels quite dramatically while enhancing the visual experience. Since the eye responds to stimuli logarithmically, a reduction of light levels by up to a factor of ten would not be noticeably different if contrast ratios are controlled. Energy consumption, however, would drop dramatically. A tenfold reduction in light consumption would produce a corresponding tenfold reduction in lighting energy use, but its real impact would be much greater. If we considered the reduction in terms of delivered energy—that is, from coal mine to lamp—then the total energy reduction would be closer to a hundredfold.

By privileging systems that produce ambient conditions in a building, not only are we significantly reducing the effective efficiency of the fundamental process (e.g., cooling the body), but we are also constraining the exploitation of natural and passive systems. We expect natural ventilation systems to provide the same level of heat dilution as mechanical ventilation, and we expect daylighting systems to maintain the same even light levels as electric lighting. Many would argue, however, that we have little other choice than to design for ambient conditions. Building interiors are so variable and occupants are so often unpredictable in their movements and activities that it would have been quite difficult to design a non-ambient lighting system that maintained acceptable performance.

Now, however, we are beginning to have the tools and technologies to light at small scales. Occupancy sensors are routinely used to shut down light systems when no one is present, and luminance and position sensors have the potential to manage the transient lighting levels in discrete locations as the sun and people move within a space. Simulation tools for energy and light analysis have already begun to simplify the lighting design process, allowing the designer to optimize the relationship between daylighting and artificial lighting as well as to determine the most suitable materials and luminaire positions. Indeed, any building owner equipped with a lighting simulation could simply reconsider the color or type of paint used in order to reduce the light necessary for maintaining appropriate contrast ratios. Small differences in either the spectral reflectance (color) or the texture (matte or gloss) of paint on adjacent surfaces can deliver contrast as effectively as a task lamp. Furthermore, many of the new technologies such as fiber optics and light-emitting diodes (LEDs), previously considered unsuitable for lighting at high levels or in large spaces, can be incorporated as we move to discrete lighting. Instead of lighting large surfaces, we could use these tiny lamps to provide light only where and when we need it. For example, fiber optics can be discretely positioned immediately adjacent to surfaces, eliminating the need for inefficient overhead lighting with its large heat gains, and LEDs allow for precise color mixing and illuminance that can be digitally configured to respond to any interior need. These new technologies bring many benefits for energy reduction: they allow for direct control of contrast, they reduce lighting losses due to the position and distance of the lamp, and they significantly reduce heat generation in a building, thereby reducing cooling costs. And perhaps of most value to the designer, a thoughtful heterogeneity of lighting could improve the aesthetics and functionality of the space while significantly reducing energy use.

The simplicity of these solutions seems to render them suspect, but the science behind this approach is sophisticated. During the past century, as ambient building systems were being disseminated, the relevant physical theories were still in their infancies. Only recently have we understood the neurological behavior of the eye. Furthermore, the simplicity is belied by its counterintuitive actions. While most architects would agree with the conclusion that reducing light levels saves energy, very few would agree with the premise that reducing light levels improves seeing and enhances the visual experience in a space. It seems to be a non sequitur. We are immensely cheered when we open our curtains to morning sunlight; we routinely turn on another lamp when unable to read easily. By thinking of light in terms of a relative contrast, and not an absolute level, we can begin to realize that our instinctual responses are dependent on the contrast, but that the contrast is also dependent on the extant light. Lowering the light level in a space with an already low contrast ratio will increase the gloom and reduce the ability to see. If the contrast level is increased as the light levels are lowered, then we will perceive the lower light space as brighter than the higher light space. This is simple to demonstrate but not so simple to imagine.

The solutions for reducing energy are perhaps not so difficult.

The line in the center is actually the same shade from top to bottom, but the dark surrounding makes it appear lighter, and the light surrounding makes it appear darker. Designers use this phenomenon to reduce the amount of light needed, such as for reading. Courtesy of the author.

Determining the right problem to solve, however, is extraordinarily complex. In the case of lighting, only physicists understand light, only psychologists understand the performance of the eye, only engineers understand electrical distribution systems, and only historians in the field of technology and culture would be aware of the pervasive role in our building choices played by marketing. Architects, however, bear full responsibility for the consequences of such systems. The need is in one discipline, but the knowledge belongs to many others.

What fundamental knowledge must we architects possess? How can we work from principles when what we do is produce artifacts? How do we take knowledge from another discipline but apply it from within ours? In the past, our approach has been one of extension— we have expanded the umbrella of our discipline to overlap with these other fields. Architectural education began to require more and more knowledge that was inherent to other disciplines, while many of these disciplines were also rapidly expanding their knowledge. The more we have extended, the more we have been forced to trade off knowledge for information. As a result, we often appropriate the knowledge from other disciplines as an ever-growing database of strategies from which we pick something that seems appropriate for the project at hand. The more complex the knowledge, the further removed it is from the architect. Design advisory tools are intended to guide the architect through a limited selection and optimization process, whereas rating systems fully remove the architect from the decision-making process. Regardless of the name—advisories, guidelines, precedents, or standards—this approach keeps us continually trapped within what is typically done.

A more recent alternative to extension is popularly termed "collaboration." The architect is still at the center of the process, but the project consultants are brought in earlier so that their knowledge can be embedded during the early design phases. Collaborative teams are most often formed when there is a definable outcome to be produced, and that outcome can range from a project, competition, or master plan to the documentation of guidelines or strategies. The focus on a definitive "artifactual" outcome of necessity tautologically bounds and limits the problem addressed by the team. Each of these consultants on the team may be representative of their discipline's knowledge, but a predefined outcome subordinates that knowledge to produce the desired solution: for instance, a sustainable building. Since disciplinary knowledge is problem based, and not solution based, this approach appropriates the other disciplines as fragments that form a whole only when they are inserted into the architectural solution. The subject of architecture is indeed larger and broader in its scope than that typically faced by the other disciplines, but it is not *supra* them.

The inherent problematic introduced by the ascension of one discipline to the center role and the relegation of other disciplines to a supporting or subordinate role will supposedly be resolved as education and research become more truly interdisciplinary. Like the terms *green* and *sustainable,* the *interdisciplinary* designation has cropped up everywhere, from the requirements in RFPs to the titles of major research centers, particularly those devoted to the environment. The aspirations for interdisciplinary work are high, for unlike in a discipline-centric approach, each discipline is a full participant in determining both the problem and the outcome. This approach is not without its structural difficulties. For an interdisciplinary team to function, there needs to be one of two types of commonalities between the disciplines: there must either be a common or shared context, or a common subject. A shared context is relatively pro forma when disciplines such as architecture and urban design work together.

Common subjects can bring many diverse disciplines together; for example, both the disciplines of public health and architecture share an interest in indoor air quality. But what happens when the disciplines are diverse enough, like architecture and physics, such that there is no natural overlapping of context or equivalent interest in a subject? Light may seem to be an area that could bring architect and physicist together, but the subjects are inherently different. The architect may be interested in improving the delivery of light in a building. The physicist may be researching how refraction in a non-homogeneous medium affects the spectral distribution of light. One wants to produce an artifact, a solution, the other wants to understand a phenomenon. How can a theory-driven profession like physics work with a practice-driven profession like architecture? It is not just the language and method that differ. In the theory-based disciplines, knowledge is becoming increasingly discrete and atomized, whereas in the practice-based disciplines, the domain is becoming larger and broader.

Since there seems to be no happy medium between focusing on sophisticated knowledge and expanding the breadth of practice, how can we mount a legitimate inquiry from within the discipline of architecture? Oddly enough, the answer may lie in broadening our scope even farther. We tend to work with multiple subjects but within a very narrow context. For example, we define our context to be a building or a group of buildings, while our subjects will range from energy to human behavior. But a building, physically, is a unit of private property, nothing more. It is not an energy system, a social system, or a political system, nor does it represent a subset of any of these systems. It is a fragment that intervenes on these larger and more complex systems. By defining our context as the building, we arbitrarily truncate it from the larger systems. An energy balance at the boundary of a building has little correlation with the building's ultimate impact on the environment.

We need to establish our context as part of the larger problem of addressing environmental impacts. By having narrowed our context to the "sustainable" building, and even more so to the "zero-energy" building, we have been doing the opposite. We have presumed that buildings are autonomous entities and that by optimizing each one with respect to its energy use we have done our part. This delimiting of the context ensures that the normative practice of architecture is unchanged, and it introduces new "solutions" in the way we most easily understand them, through products and strategies or, in essence, artifacts that can be defined and specified.

What if we expanded our context from the solution to the problem? Rather than designing more efficient lighting systems for buildings, can we ask what humans need in order to see? Rather than focusing on the design of a sustainable building, could we step back and ask ourselves how our disciplinary activities contribute to environmental degradation? Ultimately, we will come back to solutions, for inherently, as architects, we *make* things. Let us, however, think in territories larger and broader than the things we make.

2003

Notes

1. "Teaching Green," *Metropolis*, November 2002, was a report on the eponymous conference sponsored by the magazine in May 2002. One of the attendees at the conference, Vivian Loftness, suggested seven categories for evaluating universities' sustainability programs, including "the number of graduate courses and degrees related to sustainability" (14).

2. All the energy data in this article have been obtained from public docu-

ments available through the Energy Information Agency, the information branch of the Department of Energy. Among the documents cited in this article are *Annual Energy Outlook 2002 with Projections to 2020,* http:// www.eia.doe.gov/oiaf/archive/aeo02/pdf/0383(2002).pdf; and *High Performance Commercial Buildings: A Technology Roadmap* (2000), http:// www.eren.doe.gov/buildings/commercial_roadmap.

3. During the 1960s, residential living space was approximately four square meters per person. The nationwide average is now eight square meters, although in the well-developed coastal areas, such as Shanghai, the figure has already reached sixteen square meters. Government mandates are currently calling for an increase to twenty square meters per person. Lu Junhua, Peter G. Rowe, and Zhang Jie, eds., *Modern Urban Housing in China 1840–2000* (Munich: Prestel, 2001).

4. This is known as the Mexican hat effect. Retinal receptors are organized into neuron clusters (each cluster is a bipolar receptive field). The receptors in the center of the cluster are positive; that is, they send a positive signal when they intercept a photon, whereas receptors on the periphery are negative. If there is a constant light level (as such, a steady emission of photons reaching the retina), then the negative receptors will zero out the positive receptors. Note that as the visual range extends over thirteen orders of magnitude, the high and low levels of luminance that we normally encounter within buildings are relatively close in comparison. A good discussion of this effect and the neurological behavior of the eye can be found in Russell L. De Valois and Karen K. De Valois, *Spatial Vision* (New York: Oxford University Press, 1990).

5. The minimum light level is well documented throughout the literature on vision. Contrast levels can be calculated in several ways, including by simple ratios. A general rule of thumb is that minimum contrast between the illuminance of adjacent light and dark surfaces is no less than two to one.

6. General Electric published one of the earliest, and probably most thorough, of these manuals in 1942. See Charles L. Amick, *Fluorescent Lighting Manual* (New York: McGraw-Hill Book Company, 1942). Amick was an engineer with General Electric at their Nela Park facility in Cleveland. Although he claimed the manual was written for everyone—manufacturers, contractors, dealers, installers, and users—it was clearly biased toward establishing a market for the nascent lamps.

13

Here Come the Hyperaccumulators! Cleaning Toxic Sites from the Roots Up

Niall Kirkwood

Toxic trees, virus-bearing vines, plants from outer space that eat poisons for lunch, then snack on young adults. B-movies from the 1950s vividly portrayed freakish vegetation to scare audiences. In the last reel, mankind destroys the mutant greenery through the ingenuity of the scientist hero. But this is science fiction, and if the plastic ivy hanging by wires clearly visible on the movie screen is any measure, not very convincing science fiction at that. But recent eye-grabbing headlines—"Lead-Eating Mustard Plants," "Pint-Sized Plants Pack a Punch in Fight against Heavy Metals," and "Pollution-Purging Poplars"[1]—seem to have brought B-movies to life and unsettled our comfortable view of vegetation as benign and green. However, these seemingly freakish plants are in fact our very good friends.

If given the choice, plants prefer clean air, water, and soil, but of the two hundred thousand plant species, many have evolved adaptations that allow them to thrive in harsh conditions, even in the presence of extreme chemical pollution. "They do this very efficiently and without apparent harm to themselves," says Steven Rock, an environmental engineer and leading researcher with the U.S. Environmental Protection Agency's National Risk Management Research Laboratory in Cincinnati.[2] Rock spends most of his time in the laboratory and field studying these plants. Some of them can take contaminants into

Little Shop of Horrors. Film still courtesy Photofest.

their roots, stems, or leaves and use enzymes to break them down entirely or make them into new safe compounds. Some plants host microbial activity in their roots and surrounding soils, activity that accelerates the chemical breakdown of toxins.

Rock is now focusing on "hyperaccumulators," plants that have the capacity to absorb large quantities of contaminants—including heavy metals and poisonous man-made chemicals—from the surrounding water and soil.[3] He has coined the term *phyto-technology* for the use of plants for environmental cleanup.[4] Industry and federal government researchers are now at the vanguard of a broad partnership of scientists exploring ways of overcoming the toxic legacy of industrialization with safer and cheaper means than those now most often used. The plants receiving most scrutiny are poplar trees, alfalfa, sunflowers, ferns, and fescues.[5] Thirty-one academic institutions around the world, including the Harvard University Graduate School of Design, are undertaking phytoremediation research.[6] Phyto-technology is a potentially lucrative industry. David Glass, a researcher in Needham, Massachusetts, estimates that the phyto-technology market in the United States will quadruple from $50 million to $86 million in 2000 to a projected $235 million to $400 million by 2005.[7]

Phyto-technology is a new name for an old technique. Early prospectors in Europe scoured the landscape for indicator plants that signaled the presence of metal ores below the surface. One such plant was alpine pennycress, a wild perennial herb that naturally occurs in the alpine areas of central Europe as well as in the Rocky Mountains. It is the only plant that can thrive on zinc- and nickel-rich soils. The application of "phyto-mining" research took one step further in 1998 with the development by Rufus Chaney and his colleagues at the U.S. Department of Agriculture of a method to use such plants to phyto-mine nickel, cobalt, and other metals. As Chaney reported in a recent article, "Crops would be grown as hay. The plants would be cut and baled after they'd taken in enough minerals. Then they'd be burned and the ash sold as ore. Ashes of alpine pennycress grown on a high-zinc soil in Pennsylvania yielded 30 to 40 percent zinc—which is as high as high-grade ore. Electricity generated by the burning could partially offset biomining costs."[8]

The science supporting phyto-technologies is not new but reflects another way of applying knowledge about how plants and their associated microbial communities in roots and the surrounding soil treat pollutants. One limitation has been the depth to which plant roots will go to interact with subsurface pollutants and aqueous waste streams. Landscape architect Alan Christensen has recently obtained a patent for a system he developed to facilitate "deep phytoremediation," using existing landscape engineering technologies to plant poplar trees (as bare stems sprouting out their tops and sides) eight feet below the ground surface, thus facilitating horizontal rooting.[9]

When a plant encounters a contaminant in soil, water, or air, it can interact with it in four possible ways. It can find the contaminant toxic and die, can ignore it completely, can transform it into products that are useful to it, or can simply store it. On-site remediation using plants occurs in one of three ways: by *accumulation* of toxic substances in plants, by *degradation* of the pollutants by plants, or by *control* of the movement of subsurface toxic streams with trees.[10] In phyto-accumulation the leaves, shoots, and accumulated toxic substances from the top layer of soils are harvested and incinerated two to three times during the growing season.[11] In phyto-degradation, contaminants are broken down in the soil by secreted plant enzymes or by microbial activity.[12] Hydraulic control by trees such as willows and poplars uses deep rooting structures to take up and sometimes transpire through their leaves large amounts of water from the soil and groundwater.[13]

A specific example of the potential of plant uptake of contaminants is the use of Indian mustard to remove lead from the top twenty inches of contaminated soil on an abandoned former battery and pen factory site. Trenton's Gould National Battery site was home to commercial lead-acid battery manufacturers from the mid-1930s to the early 1980s. Throughout the 1980s, the property was host to a manufacturing plant for Magic Marker Industries, which abandoned the site in 1989. Lead poisoning is the nation's leading environmental risk to children under the age of six, and it is estimated that 18 percent of U.S. homes have soil lead levels above the recommended limit of 400 parts per million from lead paint, leaded gasoline, and nearby industrial facilities.

Lead does not degrade, so extraction is the only viable remediation. Indian mustard crops on test plots thirty by fifty feet reduced lead from 1,200 parts per million to below 400 parts per million in harvests between May 1997 and October 1998. This level is within residential standards set by the Environmental Protection Agency for human exposure to lead in soils. Lead levels were reduced in 75 percent of the site. Based on this test, costs of between $60,000 to $100,000 are projected to clean an acre of urban soil to a depth of twenty inches. More significant, planting and harvesting crops of Indian mustard over three growing seasons would result in 500 tons of plant biomass to be disposed of by incineration. This figure is only 0.25 percent of the total 20,000 tons of soil that would require excavation haulage (at a cost of $600,000) to the nearest hazardous landfill.

Living plants naturally able to accumulate large amounts of contaminants were first described in 1885.[14] But research on them did not start in earnest until the mid-1970s in the United States, when the basic mechanisms of phytoremediation were described in laboratory experiments. However, research into how to use and enhance the efficiency of hyperaccumulators began only a few years ago in part due to the search for a more sustainable, cheaper, and less intrusive method of cleaning contaminated sites. After the 1986 Chernobyl nuclear accident, for instance, mustard crops were planted to absorb radioactive metals such as cesium and strontium from the surrounding soils.[15] By a new aquarium building on the Cooper River in Charleston, South Carolina, the U.S. Environmental Protection Agency (EPA) planted rows of fast-growing poplar trees in 1999 to suck up cancer-causing polluted groundwater.[16] And in summer 2001, scientist Mark Ellis and landscape architect Jean Hegarty, with assistance from the Center for Technology and Environment at the Harvard University Graduate School of Design, installed perennial turf grass lawns to absorb harmful metals from the soil in four threadbare backyards in East Cambridge, Massachusetts.[17] Turf grass produces a large biomass and accumulates lead concentrations as high as 5,000 parts per million in its blades. The lead bioaccumulation factor for this variety of turf grass lawn was 9.6—it accumulates a lead concentration approximately ten times higher than that in the surrounding soil.

Ground covers, trees, and grasses may seem like unlikely tools to get rid of pollutants that have traditionally been removed with excavators, pumps, burners, washers, membranes, and injected chemicals. There is a vast array of methods to treat contaminants with heat, steam, electricity, and oxygen. The traditional "digging and dumping" of polluted soils and "pumping and treatment" of groundwater require substantial outlay of equipment, create continuous dust and noise, and need a clear separation of the cleanup from both the public and those engaged in other redevelopment activities on-site. The fact that funding for the East Cambridge program came from the EPA suggests that the national consensus on remediation technologies may be shifting away from mechanical means. Any radical shift would, however, be fiercely resisted by the billion-dollar cleanup industry, which is heavily invested in sophisticated machinery, expensive operators, and marketing. In addition, the cleanup industry is supported by regulatory conventions that define replicable best practices as those that the industries employ. Under these circumstances, can phyto-technology be a silver bullet for the environmental woes of postindustrial brownfields, or will it be just a passing novelty? Should you invest, for example, in plant pollution nurseries?

Research well funded by the EPA, the U.S. military, Dow Chemical, Chevron, and Ford, to name a few among many institutions, suggests that phyto-technology may be a more realistic and sustainable solution to the problems of industrial lands and urban brownfields than more conventional technologies. In particular, hyperaccumulators could change the way corporations, cities, and communities that own, inherit, or buy these sites might rethink the rebuilding and reuse of formerly contaminated lands. Phyto-technologies involve a synthesis of chemistry, ecology, applied horticulture, and environmental engineering leading to knowledge of what plants accumulate toxins at what speeds. Will planning and design professions profit from this innovative technology, and, more significantly, will designers use hyperaccumulation as an on-site process with rich form-making properties?

To optimize the effects of phytoremediation, plants are generally densely sown in either rows perpendicular to the flow of a subsurface groundwater plume or in clumps or patches known as "hotspots," where the contamination occurs in concentrated soil areas. The task designers now have to face is translating these applied scientific and environmental engineering techniques to the tasks of form-making and the physical realization of new site programs related to the remediation and reuse of brownfields and other derelict sites. Scientific discoveries and an understanding of their underlying patterns and structures can inspire new site design languages. The potential also exists to combine and integrate phytoremediation with other forms of environmental design and planning enterprises, for example, pump and treatment wetland systems, "brightfields" (federally funded passive solar energy areas), and soil stabilization programs.

Unfortunately, phytoremediation, in contrast with competing remedial technologies, takes a long time and relies on biological systems that are relatively unpredictable. Its low cost (approximately seventy to a hundred times cheaper) makes it appropriate for projects that do not need rushing, such as most community-based land redevelopment. Phytoremediation is effective only in certain conditions. A phytoremediation scheme will make sense only if there are appropriate growing conditions, contaminant densities, and aeration of the soil.

However, phytoremediation has enabled land reuse in several cases, and it does offer landscape architects a possible niche in the broader field of remediation. Furthermore, landscape architects and planners may play a role in the prevention of brownfields by designing sites and land-use codes that anticipate the presence of toxins and mitigate that presence with hyperaccumulators.

During the "Manufactured Sites" conference at Harvard University Graduate School of Design in 1998, Steve Rock declared, "On some sites it is possible to place planting in such a way as to allow partial reuse of the site for public access or ongoing development while the cleanup is in process. Here phyto-remediation and creative site design are united by the use of planted systems that both remediate and at the same time establish spatial and functional patterns of use." This engineer and scientist sounds the clarion: here come the hyperaccumulators!

2003

Notes

1. V. Boyd, "Pint-Sized Plants Pack a Punch in Fight against Heavy Metals," *Environmental Protection,* May 1996. M. Peterson, "Lead-Eating Mustard Plants," *New York Times,* 23 October 1996. Mia Schmiedeskamp, "Pollution-Purging Poplars, Trees That Break Down Organic Contaminants," *Scientific American,* December 1997.

2. The National Risk Management Research Laboratory of the U.S. Environmental Protection Agency Office of Research and Development "conducts research into ways to prevent and reduce risks from pollution that threaten human health and the environment." Dr. Rock works in the Remediation and Containment branch, conducting and coordinating research into phyto-technologies and is also cochair of the Phytoremediation Research Technology Development Forum. For further information see the Web site at http://www.epa.gov/ORD/NRMRL.

3. The term *hyperaccumulators* was first used in print in 1997 by R. R. Brooks et al. in "Detection of Nickeliferous Rocks by Analysis of Herbarium Specimens of Indicator Plants" in the *Journal of Geochemical Exploration* 7: 49–77. The term was used to describe plants that contain greater than 0.1 percent nickel in their dried leaves. Threshold values were then established for zinc, copper, iron, manganese, chromium, and lead. Alan Baker of University of Sheffield is currently one of the pioneers in the identification of hyperaccumulating plant species.

4. The term *phyto-technology* was introduced by Steven Rock at the "Phytoremediation, State of the Science" conference in Boston, Massachusetts, May 2000, sponsored by the U.S. Environmental Protection Agency.

5. An example is the partnership of the Phytoremediation of Organics Action Team Research (POATR) founded by the U.S. Environmental Protection Agency as part of its Remediation Technology Development Forum in 1997 with members from industry, government, and academia. These include Chevron, Dow Chemical, Goodyear, Rohm and Haas, Union Carbide, Argonne National Laboratory, the U.S. Department of Energy, and the U.S. Army. For more information on POATR, see the Web site www.rtdf .org/phyto.htm.

6. Twenty-two academic institutions are currently involved with phytoremediation research in the United States, including Rutgers, Utah, University of Massachusetts at Amherst, Kansas State, Arkansas, Oklahoma, Tennessee, University of Washington, and University of Wisconsin. Other universities involved are located in Canada, New Zealand, Britain, and Germany. Research at the Harvard University Graduate School of Design is focused on the application of phyto-technology systems in physical planning and design, particularly on postindustrial landscapes.

7. D. J. Glass, *The 1998 United States Market for Phytoremediation*

(Needham, Mass.: D. Glass Associates, April 1998, updated July 1999). See www.channel1.com/dglassassoc/INFO/phytrept.htm.

8. The work of Rufus Chaney of the U.S. Department of Agriculture at the ARS Environmental Chemistry Laboratory in Beltsville, Maryland, is described in Don Comis, "Phytoremediation: Using Plants to Clean Up Soils," *Agricultural Research,* June 2000.

9. Alan Christensen developed the early work on "deep phytoremediation" in the Harvard University Graduate School of Design seminar on advanced landscape technologies taught by the author in spring 1998. He returned to landscape practice in Utah, where he carried out field trials on his system prior to a successful patent application.

10. For an introduction to the mechanisms of phytoremediation, see Steven Rock, "Phytoremediation: Integrating Art and Engineering through Planting," chapter 5 in *Manufactured Sites: Rethinking the Post-Industrial Landscape,* ed. Niall Kirkwood (London: Spon Press, 2001).

11. Hyperaccumulators such as poplar trees, sunflowers, fescues, and ferns can absorb high levels of metals such as copper, chromium, lead, nickel, zinc, and radionuclides within their roots and concentrate them either there or in shoots and leaves. A more detailed explanation of the accumulation and harvesting method is found in the description of the cleanup at a classic brownfield site. See Robert K. Tucker and Judith Auer Shaw, "The Magic Marker Site in Trenton, New Jersey: A Case Study," in *Phytoremediation of Toxic Metals, Using Plants to Clean Up the Environment,* ed. Ilya Raskin and Burt D. Ensley (New York: John Wiley & Sons, 2000). This Trenton site is a landmark field study of phyto-technologies in which scientific and environmental purposes were linked to political action on former contaminated industrial sites in residential neighborhoods and local community health concerns.

12. Engineered plants exude specific enzymes that induce rhizospheric bacteria to degrade anthropogenic toxins. Individual plant species are also capable of destroying organic compounds in a strategy called phytostabilization. In the case of poplar trees, the trees exude a suite of enzymes that change the chemical structure of many petroleum products.

13. Phyto-technology systems are commonly planted as either crops to be harvested (transpiration ratio ranges between 850 and 300 kg of water used per kg of dry matter produced) or as blocks or lines of trees. Poplars have been observed to transpire four gallons of groundwater per tree, per day. This enables some site techniques to cleanse water above and below the water table and to control the local hydrology.

14. Baumann first described the inordinately high uptake of zinc by *Thlaspi calamnare* in 1885. Noted in B. H. Robinson, R. R. Brooks, A. W. Howes, J. H. Kirkman, and P. E. H. Gregg, "The Potential of the High-Biomass Nickel Hyperaccumulator *Berkheya coddii* for Phytoremediation

and Phytomining," *Journal of Geochemical Exploration* 60, 2 (December 1997).

15. The U.S. Department of Energy sponsored an international workshop in Slavutych, Ukraine, in 1998 to evaluate the feasibility and research needs of applying phyto-technologies to the "exclusion zone" around the Chernobyl facility. Greenhouse and field testing has started.

16. The phyto-technology project arose from the first large-scale urban application and monitoring of plant cleanup technologies in the United States at Calhoun Park, Charleston, South Carolina. The participants included the U.S. EPA Innovative Technologies Group, Cincinnati; the National Park Service; City of Charleston; South Carolina Electric Utility Company (SCEUC); The Bioengineering Group; and the Center for Technology and Environment (CTE) at the Harvard University Graduate School of Design. The project proposed to intercept a contaminated waste plume from a former manufactured gas site now owned by the SCEUC as part of an urban waterfront redevelopment project involving an aquarium, a mixed-use retail and entertainment complex, a public park, and a National Park Service ferry dock. Rows of fast-growing cottonwood trees were planted at the edge of the utility facility, and monitoring is still under way.

17. Field-testing of contaminated residential yards in Cambridge, Massachusetts, was funded by U.S. EPA FY 2000 Grant Funds for Technical Studies. A team consisting of Edenspace Systems Corporation, Virginia; Lead-Safe Cambridge, Massachusetts; and CTE, Harvard University Graduate School of Design, carried out a study titled "The Use of Phytoremediation to Minimize Human Exposure and Health Risk Following Renovation and Remodeling of Homes Containing Lead-Based Paints."

14

Neocreationism and the Illusion of Ecological Restoration

Peter Del Tredici

It is easy to enumerate hot-button issues in contemporary American culture: gun control, abortion, globalization, sprawl, climate change, gay marriage, terrorism, and illegal immigration are just a few. One thing that the debates about these issues have in common is that they are highly polarized, with neither side paying much attention to what the other is saying. Another is that they often have an overarching moralistic tone that pits absolute good against absolute evil. How our society will manage to actually move forward on these contentious issues is an unanswered question.

Within my own narrow field of expertise, plant ecology, the use of exotic versus native species in designed landscapes is an issue that seems to bring out the worst in people, not unlike the debates over gun control or abortion. As a representative of the Arnold Arboretum of Harvard University, I am a member of the Massachusetts Invasive Plant Advisory Group, a voluntary collaboration of nursery professionals, conservationists, land managers, and representatives from various government agencies that reports to the state's Executive Office of Environmental Affairs (EOEA). During the past two years, the group has produced a list of species that are invasive in "minimally managed" habitats and is in the process of developing a strategic plan for how best to deal with the problem.

Kudzu, "the vine that ate the South," along a highway near Charlotte, North Carolina. Photograph by Peter Del Tredici.

On the national scale, researchers have determined that invasive species are an ongoing threat to rare and endangered species (thus also to biodiversity) and the cause of economic losses totaling approximately $137 billion annually.[1] From the perspective of the federal government, invasive species need to be controlled both to fulfill the legal requirements of the Endangered Species Act and to save money. Proposals on the table (put forward in President Clinton's 1999 Executive Order on invasive species) recommend, among other things, the creation of rapid response "SWAT" teams that could go into the field and eradicate potential invasions while they are supposedly still manageable.[2]

Implicit in the proposals that call for the control or eradication of invasive species is the assumption that the native vegetation will return to dominance once the invasive is removed, thereby restoring the "balance of nature." That is the theory; reality is something else. Land managers and others who have to deal daily with the invasive problem know that often as not the old invasive comes back following eradication (reproducing from root sprouts or seeds), or else a new invader moves in to replace the old one. The only thing that seems to turn this dynamic around is cutting down the invasives, treating them with herbicides, and planting native species in the gaps where the invasives once were. After this, the sites require weeding of invasives for an indefinite number of years, at least until the natives are big enough to hold their ground without human assistance.[3]

What is striking about this so-called restoration process is that it looks an awful lot like *gardening*, with its ongoing need for planting and weeding. Call it what you will, but anyone who has ever worked in the garden knows that planting and weeding are endless.[4] So the question becomes: Is "landscape restoration" really just gardening dressed up with jargon to simulate ecology, or is it based on scientific theories with testable hypotheses? To put it another way: Can we put the invasive species genie back in the bottle, or are we looking at a future in which nature itself becomes a cultivated entity?[5]

The answer to this question lies in an understanding of the concept of ecological succession, the term used to describe the change in the composition of plant and animal assemblages over time. In the good old days (prior to World War II), ecologists tended to view succession as an orderly process leading to the establishment of a "climax" or steady-state community that, in the absence of disturbance, was capable of maintaining itself indefinitely. I refer to this as the Disney version of ecology, stable and predictable, with all organisms living in perfect balance. Following World War II, a younger generation of ecologists began challenging this static view, eventually formulating the theory of patch dynamics, which viewed disturbance as an integral part of a variable and unpredictable succession process.[6] The key concept here is that the nature, timing, and intensity of the disturbances are critical factors—along with climate and soil—in determining the composition of successive generations of vegetation. From the contemporary perspective, the apparent stability of current plant associations is an illusion; the only thing we know for sure is that they will be substantially different in fifty years.[7]

If one defines disturbance broadly to include human-caused disturbances such as urban sprawl, acid rain, and global warming, it becomes clear that there is no place on earth that does not experience some form of human disruption.[8] Indeed, the scary thing about huge disturbances like global warming is the uncertainty about how they will play out over the next hundred years. The absurd idea that climate change has not been "proven" grows out of the idea that people have the capacity to understand how the world actually works. When and if scientists get around to predicting precisely the effects of pumping massive amounts of carbon dioxide into the atmosphere, it will be far too late do anything about it.

To assert that planting native species will restore the balance of nature is just another way of ignoring the problem. Native plants are great, but without ongoing care and maintenance, they will die just like all the other plants we try to cultivate. The concept of implementing ecological restoration in an urban or suburban context is particularly problematic. With all that pavement, road salt, heat buildup, air pollution, and soil compaction, the urban landscape can be an inhospitable place for plants. The critical question facing landscape architects in these situations is not what plants grew there in the past but which will grow there in the future.

The successful design of urban landscapes calls for a careful analysis of the conditions that prevail on the site, followed by a determination of what species are best able to tolerate these conditions. The issue of where a given plant comes from must be secondary to the issue of its future survival. Again, the sad thing about the debate over native versus exotic species is that it has become so polarized. At its most simplistic level, native is equated with good, exotic with bad. This dichotomy ignores the fact that many plants—such as lilacs, daylilies, and hybrid rhododendrons—are neither native to eastern North America *nor* invasive. It also ignores the fact that some native species—such as poison ivy and ragweed—can be both highly invasive and highly toxic.

What I find particularly depressing about the "native species only" argument is that it ends up denying the inevitability of ecological change. Its underlying assumption is that the plant and animal communities that existed in North America before the Europeans arrived can and should be preserved. The fact that this pre-Columbian environment no longer exists—and cannot be re-created—does not seem to matter.[9] Many landscape professionals have a strong desire to restore habitats to the way they used to be, even after the original conditions that produced these assemblages of plants and animals have long since disappeared. To deny the inevitability of ecological change or to pass moral judgment on it is to deny the reality of organic evolution.[10]

The fact is that plants from around the world have been brought together in our cities, and some have flourished and reproduced. In many cases, these plants are actually performing significant ecological functions, such as absorbing excess nutrients from polluted waterways. A good example is the common reed, *Phragmites australis,* which is native to both Europe and North America and occupies brackish wetlands up and down the East Coast, most dramatically in the meadowlands bordering the New Jersey Turnpike west of Manhattan. While *Phragmites* is often portrayed as the ultimate invasive species, it is actually mitigating pollution in such sites by absorbing a great deal of nitrogen and phosphorus from the environment.[11] From the functional perspective, invasive species can be viewed as a symptom of environmental degradation rather than a cause.

Regardless of how one feels about the unique assemblages of plants that grow in our cities, they are the forests, fields, and wetlands of the future, and their diversity and spontaneity mirror that of the society at large. Indeed, the very same processes that have led to the globalization of the world economy—unfettered trade and travel among nations—have also led to the globalization of our environment. The main difference between the two, however, is that the environment is more complicated and harder to control than the economy.

So what can landscape architects, designers, and contractors do about these impending changes? My advice is simple: do not limit your planting designs to a palette of native species that might once have grown on the site. Imposing such a limitation on diversity not only reduces the aesthetic possibilities for the landscape but also its overall adaptability. As a graceful way out of the native versus exotic debate, I recommend using sustainability as the standard for deciding what to plant. According to my definition, sustainable landscape plants can tolerate the conditions that prevail on the site, require minimal applications of pesticides, herbicides, and fertilizers to look good, have greater drought tolerance and winter hardiness than other plants, and do not spread aggressively into surrounding natural areas. From this perspective, *invasiveness* is but one of several criteria that should be used when selecting plants for a given site, and *sustainability* means that the final planting list is based on a realistic evaluation of site conditions rather than on a romantic notion of the past.

2004

Notes

1. David Pimentel, Lori Lach, Rodolfo Zuniga, and Doug Morrison, "Environmental and Economic Costs of Nonindigenous Species in the United States," *BioScience* 50, 1 (2000): 53–65.

2. William J. Clinton, "Invasive Species," Executive Order 13112, February 3, 1999, *Weekly Compilation of Presidential Documents* 35, 5 (February 8, 1999): 157–210.

3. In 2003, the Limahuli Garden on the island of Kauai in the Hawaiian

Islands was spending approximately $30,000 per acre on removing invasive species and replanting native vegetation. How successful this treatment will be remains to be seen.

4. Peter Del Tredici, "Nature Abhors a Garden," *Pacific Horticulture* 62, 3 (2001): 5–6.

5. Daniel Janzen, "Gardenification of Wildland Nature and the Human Footprint," *Science* 279 (1998): 1312–13.

6. Michael G. Barbour, "Ecological Fragmentation in the Fifties," in *Uncommon Ground,* ed. William Cronin (New York: W. W. Norton, 1995), 233–55.

7. Jean Fike and William A. Niering, "Four Decades of Old Field Vegetation Development and the Role of *Celastrus orbiculatus* in the Northeastern United States," *Journal of Vegetation Science* 10 (1999): 483–92.

8. See Bill McKibben, *The End of Nature* (New York: Random House, 1989).

9. Shepard Krech III, *The Ecological Indian: Myth and History* (New York: W. W. Norton, 1999).

10. Steven J. Gould, "An Evolutionary Perspective on Strengths, Fallacies, and Confusions in the Concept of Native Plants," *Arnoldia* 58 (1998): 2–10.

11. Craig S. Campbell and Michael H. Ogden, *Constructed Wetlands in the Sustainable Landscape* (New York: John Wiley and Sons, 1999).

15

A Word for Landscape Architecture
John Beardsley

I wish to speak a word for landscape architecture, for design inextricable from the history of a site, from its spatial, material, and phenomenal conditions, and from natural and social ecology, as contrasted with a design merely of buildings—to regard design as a part and parcel of nature, as well as of society. I wish to make an extreme statement, if only to make an emphatic one, for there are enough champions of architecture.

Anyone familiar with "Walking," by Thoreau, will recognize that I have borrowed the rhetoric of the preamble of his essay. Thoreau used hyperbole to make a point; I am inclined to do the same in order to argue that landscape architecture will soon become the most consequential of the design arts. Admittedly, the profession has been beset by various problems. Relatively young, it lacks the rich theoretical and critical traditions of architecture. It has long been constrained by an attachment to the picturesque. In recent years it has been at war within itself, diverse factions pitting ecology against art—as if the two could not coexist. And so far it has failed to attain the public profile of architecture or the fine arts: built works of landscape architecture are not as readily identified and evaluated as paintings, sculptures, or buildings.

Yet as the landscape architect Laurie Olin has written, "It is hard to think of any field that has accomplished so much for society with so few people and with so little understanding of its scope or ambitions."[1]

Much in the history of the discipline substantiates this large claim: the nineteenth-century parks that enhance so many American cities; the national and state park systems; the rise of urban planning in the 1920s, which was an outgrowth of landscape architecture; the development of prototypical garden cities; the stunning Modernist works of such designers as Daniel Urban Kiley, James C. Rose, and Lawrence Halprin; and the embrace of ecology in recent years as a moral compass for the profession. Pressing Olin's argument, I would insist that recent achievements in landscape architecture are as visually compelling as those of the past and even more technically sophisticated and conceptually complex. Combining elements of architecture and sculpture with knowledge from the natural sciences, landscape architecture today is struggling to meet profound environmental, social, technological, and artistic challenges.

Thirty years ago, in *Design on the Land*, historian Norman Newton could confidently describe landscape architecture as "the art—or the science if preferred—of arranging land, together with the spaces and objects upon it, for safe, efficient, healthful, pleasant human use."[2] This definition is too terse for the intricacies of practice today. We are now apt to view landscape architecture as an "expanded field," as a discipline bridging science and art, mediating between nature and culture.[3]

Landscape architecture is neither art nor science, but art *and* science; it fuses environmental design with biological and cultural ecology. Landscape architecture aims to do more than to produce places for safe, healthful, and pleasant use; it has become a forum for the articulation and enactment of individual and societal attitudes toward nature. Landscape architecture lies at the intersection of personal and collective experiences of nature; it addresses the material and historical aspects of landscape even as it explores nature's more poetic, even mythological, associations.

Complexity alone cannot engender consequential works of art. Significant cultural expressions often result from the convergence of a compelling artistic language with an urgent external stimulus. The rise of Cubism, for instance, can be viewed as a register of the radical social and technological transformations of early-twentieth-century modernization, just as the emergence of Surrealism can be seen as an expression of the influence of Freudian theory. The consequences of such convergences are discernable in design as well as in art. Urgent external stimuli have lately been much in evidence in landscape architecture. Demands for the restoration of derelict and often toxic industrial sites pose artistic, social, and technical difficulties; so does the need to reuse abandoned sites in declining urban centers. The emergence of environmentalism and the ethic of sustainable design are encouraging the development of "green" infrastructure for improved energy efficiency, stormwater management, wastewater treatment, bioremediation, vegetal roofing, and recycling. Intensifying suburban and exurban sprawl requires new strategies for landscape management and open-space preservation. Continued population growth, especially in the Third World, is heightening the need to develop minimum standards for the provision of urban green space, while increased leisure time in the developed world is placing unprecedented burdens on parks and other natural places of recreation. Landscape practitioners today are grappling as well with the dilemma of designing at radically different scales—from that of the small urban space to that of the entire ecosystem.

These phenomena raise an important question. Are these urgent social and environmental demands being met by the development of a compelling design language—a language particular to landscape architecture? Landscape architect Diana Balmori has articulated widespread anxieties within the profession that landscape architecture has yet to find a contemporary idiom. "The profession of landscape architecture appears to be finished," she argued. "Its edges have been overtaken by architects and environmental artists. Ecology has been taken over by engineers and hasn't really affected design. At the same time, the profession hasn't found a core. The center has not been defined and held."[4]

In my view, the situation is not nearly so dire. I would argue that external pressures and contemporary expressive means are indeed working together in recent landscape architecture. I would argue too that this convergence is providing the profession with compelling narratives that might restore the sense of a vital center and help it achieve the visibility so lacking in recent decades.

One such narrative is sustainability—an idea that increasingly informs the design of buildings and landscapes. Indeed, sustainable design seems to require some degree of interdisciplinary cooperation or hybridization in the creation of "green" infrastructure. Examples of this kind of work include landscape designer Herbert Dreiseitl's unusual and visible stormwater retention and purification system for architect Renzo Piano's DaimlerChrysler complex at Potsdamer Platz in Berlin. This ambitious scheme features rooftop gardens that capture and filter rainwater, which is then directed to cisterns and used in the building. The cisterns also feed a large lagoon, where reeds provide physical and biochemical cleansing; mechanical filters furnish backup purification. The benefits of this scheme are not only technical but also aesthetic, even educational; not simply an element of infrastructure, the lagoon is an attractive public amenity that offers lessons in and demonstrations of urban hydrology. More generally, such collaborations suggest that the knowledge provided by landscape architects is increasingly essential to the responsible practice of architecture.

Landscape remediation is another narrative resulting from the convergence of contemporary subject and idiom. At the two-hundred-hectare site of the former iron and steel plant Duisburg-Meiderich in Germany, the landscape architecture firm of Latz + Partner has designed a park that sets new standards for reclamation—for the kind of reclamation, moreover, that does not disguise the problematic history of its site. Designed by Peter Latz and Anneliese Latz, Landscape Park Duisburg North makes an aesthetic and political strategy out of revealing site disturbances. The facility, abandoned by Thyssen Steel in 1985, included blast furnaces, ore bunkers, and a sintering plant; it was crisscrossed by roads, rail lines, and a canal. The soil of the site was contaminated with heavy metals, the canal polluted. The design of the reclamation was guided by existing infrastructure: elevated rail lines and ground-level roads were retained to provide both topographical interest and a framework for circulation. A sewer line and treatment plant were built to clean the old canal; a new stormwater collection system filled the former cooling and settling tanks—once contaminated with arsenic—with fresh water.

At the heart of the project are the preserved blast furnaces. Like other relics of heavy industry, these structures seem at once terrible and awe inspiring. To emphasize these qualities, Latz + Partner have surrounded the furnaces with trees, making them appear like craggy mountains glimpsed through a forest. There is a precedent for such industrial archaeology—I am thinking chiefly of Gasworks Park in Seattle. Both might be understood as instances of the "industrial sublime." The constructions at Duisburg surround a space that has been named Piazza Metallica, where forty-nine salvaged hematite slabs—

Atelier Dreiseitl, Urban Waters at Potsdamer Platz, Berlin, Germany; landscape design for DaimlerChrysler, Potsdamer Platz; architecture by Renzo Piano Building Workshop; main basin. Courtesy Atelier Dreiseitl.

each about 2.5 meters square and weighing nearly eight tons—are set out in a grid; the recycled metal, which once lined casting pits in the foundry, commemorates the melting and pouring processes that occurred there. Near the blast furnaces are the remains of ore bunkers that have been transformed into enclosed gardens. Deep within thick concrete walls, these gardens produce a kind of uncanny

Latz + Partner, Landscape Park Duisburg North, 1991–2000. Courtesy Latz + Partner. Photograph by Christa Panick.

juxtaposition: they are cloistered, almost monastic spaces, yet set in a menacing industrial frame. Perhaps more than any other element in the park, they convey the designers' strategy of retaining and revealing yet simultaneously transforming the structures of the site.

Several remediation techniques have been employed at Duisburg, depending on site conditions. The most toxic remnants, including the old sintering plant, were dynamited and buried. Elsewhere contaminated materials were left in place. Several large slag heaps with low-level hydrocarbon pollution, already in stable condition and colonized by plants, were left undisturbed. They are available for limited access and use while they are gradually decontaminated through bioremediation. Retaining the piles has two advantages: it prevents further dispersal of the pollutants, and it creates compelling memorials to site disturbance. Such "found" landscapes produce some of the park's most intriguing images, such as that of an ore cart—still in its tracks—sprouting vegetation.

Just as important, although less obvious, Landscape Park Duisburg North is an example of social as well as environmental restoration. A place that no longer had any real value to society and that otherwise would have been an eyesore has been given an entirely new life, one that few might have imagined it could have. In a region with little open space, the park offers significant and unusual opportunities for recreation: the blast furnace can be climbed to a height of about fifty meters; the cooling tanks are used for swimming, the concrete chimneys for climbing. At a more speculative level, the park offers a lesson in the environmental costs of modern industrial policies and an occasion to wonder about future appropriate choices. Effacing the site's history and erasing its contradictions would have been far less compelling. As Peter Latz says, "The point is, where is the imagination most challenged, in a state of harmony or in a state of disharmony? Disharmony produces a different statement, a different harmony, a different reconciliation. . . . The seemingly chance results of human interference, which are generally judged to be negative, also have immensely exciting, positive aspects."[5]

In such circumstances the role of the designer is to decide what to retain, what to transform, and what to replace. Disharmony, discontinuity, contradiction: these are the conditions driving the development of a contemporary language of landscape architecture.

But if I had to choose one project that suggests all the intricacies of recent landscape architecture, it would be Parque Ecológico

Xochimilco in Mexico City. Not only does this project articulate the commanding narratives that undergird recent practice, such as remediation and sustainability, it also addresses the challenges of urbanization in one of the most populous cities in the developing world, providing both open space for recreation and productive land for economic development. And it does all this on multiple scales, from the circulation in a flower market to the workings of an extensive ecosystem. Even more than Landscape Park Duisburg North, Xochimilco suggests the large role that landscape architecture can now play in social and environmental remediation.

Xochimilco, meaning "the place where flowers are grown," is a fragment of a pre-conquest—in places even pre-Aztec—landscape of artificial garden islands created in the lake that once filled a large area of the valley of Mexico. The islands, called *chinampas,* were constructed by piling soil on reed mats and anchoring their edges with salix trees. Dating to the tenth century, this landscape of canals and rectangular islands was declared a UNESCO World Heritage site in 1987; the designation prompted a large-scale environmental restoration project undertaken by Mexico City and the borough of Xochimilco. Designed by Mario Schjetnan of Grupo de Diseño Urbano and built in the early 1990s, the project encompasses some three thousand hectares of surviving islands.

The site presented extraordinary challenges. Many of the islands were sinking due to the many wells that fed upon aquifers. Urban development was increasing stormwater runoff and subjecting the area to increased flooding. Surface water was contaminated; canals were choked with aquatic plants. Those islands deep in the canal system were hard to reach and thus unavailable for agriculture; those nearer the edges of the site were being encroached on by unauthorized residential buildings. The design was guided by hydraulic strategies: water was pumped back into the aquifer to stabilize the site; large reservoirs were created to retain stormwater; polluted water was processed at treatment plants, and the treated water was discharged back into the lake to regulate the water levels in the canals. Eroded islands were re-created using meshes of logs filled with dredge and stabilized by salix trees. (More than one million trees were planted on the site.) Agriculture was reintroduced: some islands have pastures for grazing; others are planted with flowers and vegetables. A tree nursery was also located on the site; every year it produces thirty million trees that are then planted throughout Mexico City. Canals were cleared of harmful vegetation and rehabilitated for recreation as well as agriculture. Today, pole barges ply the canals of Xochimilco, especially on weekends; gondolas and gondoliers are available for hire at embarcaderos built along the edge of the site. Out in the canals, you can collect sustenance for body and soul: kitchen barges sell food, while others ferry professional musicians, available to serenade visitors with patriotic and romantic songs.

At one edge of the *chinampas* landscape is a three-hundred-hectare park, whose different zones emphasize natural, recreational, and interpretive areas. Water again provides the basis for design: the terraced entry is focused on imposing stone-lined aqueducts that discharge cleansed water into the lake; a plaza features a water tower in the form of an Archimedes screw. A visitor center completes the complex. It includes an auditorium and galleries with exhibitions focused on the region's ecology, archaeology, and agriculture; a roof terrace affords vistas over the lakes and canals toward distant snow-covered volcanoes. From the entry, a four-hundred-meter pergola leads to an embarcadero, past an arboretum and flower beds representing the productive activities dispersed across the *chinampas*. The remaining park area features playing fields and ball courts, wetlands to collect stormwater runoff, and demonstration agricultural zones. To enhance economic activity on the site, the largest flower market in Mexico City was built adjacent to the main highway approach. Its 1,800 stalls are fully leased and very busy, especially on weekends. In all, the park is a microcosm of the larger landscape, highlighting its ecological, historic, agricultural, and recreational attributes. More than something just to look at, this is a working landscape.

Both Landscape Park Duisburg North and Parque Ecológico Xochimilco exhibit the high ambition and conceptual complexity of contemporary landscape architecture. Each uses the history of its site to create stirring places and compelling cultural narratives. Each envisions landscape as both natural and social space embodying the potential of design to enhance cultural and biophysical phenomena. Both reveal the capacity of landscape architecture to address the challenges of degraded landscapes and to achieve at least some level of sustainability. And both are works of art; they attain a kind of iconic power in their revelation of the problems and the possibilities of the contemporary landscape.

But what is the narrative being told in these landscapes—told in the language of landscape architecture as landscape architecture? At the risk of sounding like the boor in *The Graduate,* who summed up the hero's career possibilities with the word *plastics,* I would like to respond with a single word: what is being explored and revealed in these contemporary landscapes is *entropy.* Entropy is disorder or randomness in a system. In thermodynamics, entropy measures the quantity of thermal energy, or heat, available for useful work: the greater the entropy, the less the available energy. According to the second law of thermodynamics—the law pertinent to my argument—the change in entropy of a system during any process is either zero or positive; that is to say, the amount of disorder in an isolated system is always stable or increasing. Shuffle a deck of cards, and the result will be as or more random than the initial sequence; the cards will not organize themselves into suits or into numerical order. Heat flows only from a hotter substance to a colder one, never the reverse. Gas expands to fill its container; it will not contract. As heat is dispersed or as gas expands, entropy increases. Natural processes result in a universe of greater entropy.

Those conversant with the language of contemporary art know that entropy was a particular preoccupation of Robert Smithson. Several of his earthworks can be interpreted as pedagogic exercises in entropy. Smithson dumped asphalt into a quarry and let it run randomly down a slope; he piled dirt on the roof of a woodshed until the supporting beams cracked. His *Nonsites*—sculptures created by collecting materials from a place, sorting them into bins, and exhibiting them along with maps or photographs of their sites—might be described as efforts to reverse the effects of entropy, if only temporarily. "The fact remains," he insisted, "that the mind and things of certain artists are not 'unities' but things in a state of arrested disruption."[6]

On a broadly theoretical level, Smithson recognized that patterns of human and natural disturbances in the landscape were undermining the reassuring conventions with which landscape has been represented. "The 'pastoral,' it seems, is outmoded. The gardens of history are being replaced by sites of time"—by sites, that is, that manifest the transformative effects of human action or of natural processes like erosion and decomposition. "A sense of chaotic planning engulfs site after site . . . but to what purpose?" Smithson asked. To Smithson, the struggle against chaos was enormously intriguing. Anticipating Peter Latz's argument that discord excites the imagination more than harmony, he wrote: "A bleached and fractured world surrounds the artist. To organize this mess of corrosion into patterns, grids, and subdivisions is an esthetic process that has scarcely been touched."[7]

How is entropy relevant to landscape projects like Duisburg or Xochimilco? In each project, the designers addressed conditions that were highly entropic. At Xochimilco, the islands were sinking, the soil eroding. Water was polluted; the land was unproductive, the edges of the site compromised by chaotic urbanization. At Duisburg, the steelworks had been demobilized, the energy removed from the site in a literal way; what remained was contaminated earth, polluted water, and abandoned infrastructure. Both Xochimilco and Duisburg might be interpreted as excursions into the "bleached and fractured world" described by Smithson, as efforts to hold the line, albeit briefly, against the drift toward randomness and disorder. (Entropy still marks Xochimilco: the park has not been maintained as it should be.)

All designed landscapes can be seen, in some ways, as expressions of the entropic passage of time. Here too one can find resonances between landscape and contemporary sculpture. The work of Richard Serra, especially, has been studied for how it encourages, even impels, motion through space and over time as a condition of its perception. But the time of sculpture, usually, is limited to the perceptual experience. The time of landscape architecture is more complex. No place is a tabula rasa, without history; any intervention by any designer is part of a series of interventions, of marks already inscribed or yet to be inscribed on the site. Every design is subject to the actions of dynamic and unpredictable natural and cultural forces—the continual transformations produced by growth and decay, for example, or by changing patterns of social use and habitation. Smithson once described a park not as a "thing-in-itself" but as "a process of ongoing relationships existing in a physical region"—an idea now exemplified by much forceful design work.[8]

Michael Van Valkenburgh Associates' Mill Race Park in Columbus, Indiana, and the firm's Allegheny Riverfront Park in Pittsburgh are designed to withstand periodic flooding from nearby rivers; George Hargreaves describes his firm's park projects as "theaters of the environment," intended to reveal geophysical, biological, and cultural forces at work in the landscape. If much painting and sculpture aspire to what critic and historian Michael Fried once called "presentness"—the condition of being fully observable in an instant—and if architecture strives for some measure of permanence, then landscape architecture, in contrast, struggles to embrace the dynamic.[9]

Of course, entropy is only one of many forces at work in the world. Current scientific studies of complexity propose that there may be some counterforce to the second law of thermodynamics, exemplified in the tendency of matter and biological life toward ever-greater levels of organization. Many natural systems are aptly described as chaotic—the weather, the flow of turbulent fluids, the orbit of particles—and in such systems, small changes in initial circumstances can produce big differences in subsequent conditions. But complex systems seem to change within predictable limits and to exhibit tendencies toward self-similar patterns, or fractals. Thus, the temperatures at a given place on the globe will vary but within predictable limits; clouds and waves will resemble each other but will not be exactly the same.[10]

Complexity science attempts to describe such patterns. It depicts a world that is dynamic and mutable but self-organizing at ever finer levels, for instance in the emergence of life from inert matter, in the evolution of more elaborate life forms from simpler ones, and in the increasingly intricate interdependencies within complex ecosystems like coral reefs and rain forests. Complexity theory might serve as a useful metaphor for contemporary cultural practice.[11]

Complexity is not necessarily better, but it increasingly characterizes our environmental and social circumstances. An appreciation of complexity might make cultural responses more discriminating, more robust. Landscape architecture is today exhibiting, in its own way, the tendency toward greater organization and complexity described by theorists and scientists, and in so doing it is endeavoring to keep at bay randomness and disorder. And it is this tension—between order and disorder, between organization and entropy—that provides much of the narrative power of contemporary landscape architecture.

Long overshadowed by architecture and the fine arts, landscape architecture is producing remarkable transformations in our public environments. The profession is maturing; conceptually, it is more complex. It is developing the artistic and technical tools to address extraordinary social and environmental demands. The ways in which we understand and represent our relationship with nature are enormously important in the expression of culture. The ways in which we meet the challenges of urban sprawl, open-space preservation, resource consumption and waste, and environmental protection and restoration are crucial to the quality of our lives—maybe even to the survival of our species. It is landscape architecture that confronts these challenges. I wish to make an extreme statement, if only to make an emphatic one: landscape architecture will prove the most consequential art of our time.

2000

Notes

1. Laurie Olin, letter to the Graham Foundation, August 3, 1999.

2. Norman Newton, *Design on the Land* (Cambridge, Mass.: Belknap Press of Harvard University Press, 1971), xxi.

3. Borrowed from linguistics, the term *expanded field* entered the language of contemporary art criticism chiefly through an essay by Rosalind Krauss, "Sculpture in the Expanded Field," originally published in *October* in 1978 and reprinted in her book *The Originality of the Avant-Garde and Other Modernist Myths* (Cambridge, Mass.: MIT Press, 1985), 276–90; the term was subsequently used by Elizabeth Meyer in "The Expanded Field of Landscape Architecture," in *Ecological Design and Planning*, ed. G. Thompson and F. Steiner (New York: John Wiley & Sons, 1997), 45–79.

4. Diana Balmori, in conversation with the author, January 1998.

5. Peter Latz, in Udo Weilacher, *Between Landscape Architecture and Land Art* (Basel: Birkhäuser, 1999), 129.

6. Robert Smithson, "A Sedimentation of the Mind: Earth Projects," in *The Writings of Robert Smithson*, ed. Nancy Holt (New York: New York University Press, 1979), 87.

7. Ibid., 85, 83, 82.

8. Robert Smithson, "Frederick Law Olmsted and the Dialectical Landscape," in *Writings*, ed. Holt, 119.

9. Michael Fried, "Art and Objecthood," *ArtForum*, June 1967.

10. General introductions to complexity science include M. Mitchell Waldrop, *Complexity: The Emerging Science at the Edge of Order and Chaos* (New York: Simon and Schuster, 1992); and Roger Lewin, *Complexity: Life on the Edge of Chaos* (London: J. M. Dent, 1993).

11. Charles Jencks has begun to discern the relationship between complexity science and contemporary design. See his *The Architecture of the Jumping Universe* (London: Academy Editions, 1995).

Contributors

Michelle Addington is associate professor of architecture at Yale University. Both an architect and an engineer, she teaches and researches in the areas of energy and advanced technologies. She is coauthor of *Smart Materials and Technologies for the Architecture and Design Professions.*

John Beardsley is a writer, curator, and senior lecturer in landscape architecture at the Harvard University Graduate School of Design. His books include *Earthworks and Beyond: Contemporary Art in the Landscape.*

Albert Borgmann is Regents Professor of Philosophy at the University of Montana at Missoula. His latest book is *Real American Ethics: Taking Responsibility for Our Country.*

Peter Buchanan has worked as an architect and planner in southern Africa, Europe, and the Middle East. He was deputy editor of *The Architectural Review* for a decade. Since then he has written the four-volume *Renzo Piano Building Workshop: Complete Works* and was curator of the exhibition *Ten Shades of Green.*

Peter Del Tredici is senior research scientist at Arnold Arboretum of Harvard University. He has taught at Harvard's Graduate School of Design since 1992.

Robert France is adjunct associate professor of landscape ecology at the Harvard University Graduate School of Design. His many books include *Wetland Design: Principles and Practices for Landscape Architects and Land-Use Planners, Deep Immersion: The Experience of Water,* and *Profitably Soaked: Thoreau's Engagement with Water.*

Susannah Hagan is head of the MA program in architecture and sustainability at the University of East London. Her publications include *Taking Shape: A Cultural Assessment of Environmental Architecture, City Fights* (with Mark Hewitt), and *More with Less: MCA Architects,* as well as essays on the idea of "the sustainable city."

Kristina Hill is associate professor and director of the Program in Landscape Architecture at the University of Virginia. She is coeditor of *Ecology and Design.*

Catherine Howett is professor emerita in the School of Environmental Design of the University of Georgia, a senior fellow in Studies in Landscape Architecture at Dumbarton Oaks, and author of *A World of Her Own Making: The Story of Katherine Smith Reynolds and the Landscape at Reynolda.*

Niall Kirkwood is professor of landscape architecture and technology, chair of the Department of Landscape Architecture, and founder and director of the Center for Technology and Environment at Harvard University Graduate School of Design. He is author of *Weathering and Durability in Landscape Architecture* and *The Art of Landscape Detail.*

Lucy R. Lippard is former art critic for *Art in America, The Village Voice,* and *Z Magazine.* She has written eighteen books on subjects ranging from pop art to Native American art, including *On the Beaten Track: Tourism, Art, and Place.*

Bill McKibben is the author of many books, including *The End of Nature, Hope, Human and Wild: True Stories of Living Lightly on the Earth,* and most recently *Deep Economy: The Wealth of Communities and the Durable Future.*

Michael Pollan is editor-at-large of *Harper's Magazine*. His many books include *The Omnivore's Dilemma: A Natural History of Four Meals, Second Nature: A Gardener's Education,* and *A Place of My Own: The Education of an Amateur Builder.*

William S. Saunders is editor of *Harvard Design Magazine* and author of books including *Modern Architecture: Photography by Ezra Stoller.*

Robert L. Thayer Jr., FASLA, is emeritus professor of landscape architecture and founder of the landscape architecture program at the University of California, Davis. He is author of *Gray World, Green Heart: Technology, Nature, and the Sustainable Landscape* and *LifePlace: Bioregional Thought and Practice.* As a professional he works on regenerative systems in landscape architecture, sustainable design, bioregional theory and practice, wind/renewable energy policy, and post oil-peak landscape planning.

Rossana Vaccarino is principal of Vaccarino Associates, St. Thomas, Virgin Islands. She is author of *Roberto Burle Marx: Landscapes Reflected.* She is a former assistant professor of landscape architecture at the Harvard University Graduate School of Design.

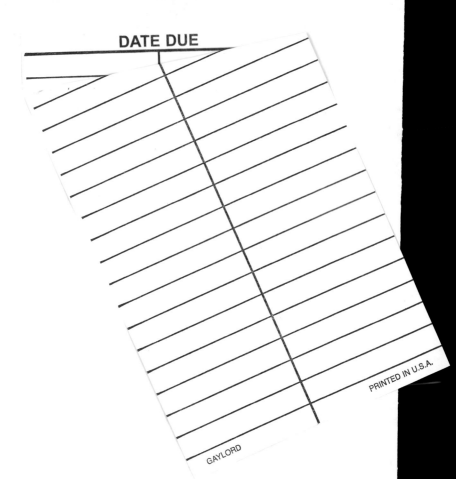

DATE DUE

GAYLORD

PRINTED IN U.S.A.